THE CRAYFISH

THE CRAYFISH

AN INTRODUCTION TO THE STUDY OF ZOOLOGY

by

T. H. HUXLEY, F.R.S.

WITH EIGHTY-TWO ILLUSTRATIONS

GYAN

Published by

Gyan Publishing House
5, Ansari Road
Daryaganj, New Delhi-110002
Phone: 011-47034999, 9811692060
E-mail: books@gyanbooks.com

Distribution Network
gyanbooks.com
India, USA, Canada, UK, Australia, France

ISBN : 978-81-212-6099-2 (PB)
First Published, 1880

2nd Impression 2021

Printed at: Gyan Press, Delhi.

The Crayfish
Author: T. H. Huxley

THE INTERNATIONAL SCIENTIFIC SERIES.

THE

CRAYFISH.

AN INTRODUCTION TO THE STUDY OF ZOOLOGY.

BY

T. H. HUXLEY, F.R.S.

WITH EIGHTY-TWO ILLUSTRATIONS.

1880.

PREFACE.

In writing this book about Crayfishes it has not been my intention to compose a zoological monograph on that group of animals. Such a work, to be worthy of the name, would require the devotion of years of patient study to a mass of materials collected from many parts of the world. Nor has it been my ambition to write a treatise upon our English crayfish, which should in any way provoke comparison with the memorable labours of Lyonet, Bojanus, or Strauss Durckheim, upon the willow caterpillar, the tortoise, and the cockchafer. What I have had in view is a much humbler, though perhaps, in the present state of science, not less useful object. I have desired, in fact, to show how the careful study of one of the commonest and most insignificant of animals, leads us, step by step, from every-day knowledge to the widest generalizations and the most difficult problems of zoology; and, indeed, of biological science in general.

It is for this reason that I have termed the book an "Introduction to Zoology." For, whoever will follow its pages, crayfish in hand, and will try to verify for himself the statements which it contains, will find himself brought face to face with all the great zoological questions which excite so lively an interest at the present day; he will understand the method by which alone we can hope to attain to satisfactory answers of these questions; and, finally, he will appreciate the justice of Diderot's remark, "Il faut être profond dans l'art ou dans la science pour en bien posséder les éléments."

And these benefits will accrue to the student whatever shortcomings and errors in the work itself may be made apparent by the process of verification. "Common and lowly as most may think the crayfish," well says Roesel von Rosenhof, "it is yet so full of wonders that the greatest naturalist may be puzzled to give a clear account of it." But only the broad facts of the case are of fundamental importance; and, so far as these are concerned, I venture to hope that no error has slipped into my statement of them. As for the details, it must be remembered, not only that some omission or mistake is almost unavoidable, but that new lights come with new methods of investigation; and that better modes of statement follow upon the improvement of our general views introduced by the gradual widening of our knowledge.

I sincerely hope that such amplifications and rectifications may speedily abound; and that this sketch may be the means of directing the attention of observers in all parts of the world to the crayfishes. Combined efforts will soon furnish the answers to many questions which a single worker can merely state; and, by completing the history of one group of animals, secure the foundation of the whole of biological science.

In the Appendix, I have added a few notes respecting points of detail with which I thought it unnecessary to burden the text; and, under the head of Bibliography, I have given some references to the literature of the subject which may be useful to those who wish to follow it out more fully.

I am indebted to Mr. T. J. Parker, demonstrator of my biological class, for several anatomical drawings; and for valuable aid in supervising the execution of the woodcuts, and in seeing the work through the press.

Mr. Cooper has had charge of the illustrations, and I am indebted to him and to Mr. Coombs, the accurate and skilful draughtsman to whom the more difficult subjects were entrusted, for such excellent specimens of xylographic art as the figures of the Crab, Lobster, Rock Lobster, and Norway Lobster.

T. H. H.

London,
November, 1879.

CONTENTS.

PAGE

PREFACE v

LIST OF WOODCUTS xi

CHAPTER I.

THE NATURAL HISTORY OF THE COMMON CRAYFISH . . . 1

CHAPTER II.

THE PHYSIOLOGY OF THE COMMON CRAYFISH. THE MECHANISM BY WHICH THE PARTS OF THE LIVING ENGINE ARE SUPPLIED WITH THE MATERIALS NECESSARY FOR THEIR MAINTENANCE AND GROWTH 46

CHAPTER III.

THE PHYSIOLOGY OF THE COMMON CRAYFISH. THE MECHANISM BY WHICH THE LIVING ORGANISM ADJUSTS ITSELF TO SURROUNDING CONDITIONS AND REPRODUCES ITSELF 87

CHAPTER IV.

THE MORPHOLOGY OF THE COMMON CRAYFISH. THE STRUCTURE AND THE DEVELOPMENT OF THE INDIVIDUAL 137

CHAPTER V.

PAGE

THE COMPARATIVE MORPHOLOGY OF THE CRAYFISH. THE STRUCTURE AND THE DEVELOPMENT OF THE CRAYFISH COMPARED WITH THOSE OF OTHER LIVING BEINGS 227

CHAPTER VI.

THE DISTRIBUTION AND THE ÆTIOLOGY OF THE CRAYFISHES . . 288

NOTES 347

BIBLIOGRAPHY 357

INDEX 363

LIST OF WOODCUTS.

PAGE

Frontispiece. THE COMMON CRAYFISH, *Astacus fluviatilis*, (MALE)

FIG. 1.	*Astacus fluviatilis.*	SIDE VIEW OF THE MALE. . .	6
,, 2.	,, ,,	DORSAL VIEWS OF MALE AND FEMALE	18
,, 3.	,, ,,	VENTRAL VIEWS OF MALE AND FEMALE	21
,, 4.	,, ,,	THE GILLS	26
,, 5.	,, ,,	DISSECTION FROM THE DORSAL SIDE (MALE)	28
,, 6.	,, ,,	LONGITUDINAL VERTICAL SECTION OF THE ALIMENTARY CANAL . .	29
,, 7.	,, ,,	A GASTROLITH OR "CRAB'S EYE" .	30
,, 8.	,, ,,	ATTACHMENT OF YOUNG TO SWIMMERET OF MOTHER . . .	41
,, 9.	,, ,,	STRUCTURE OF THE STOMACH . .	53
,, 10.	,, ,,	LONGITUDINAL SECTION OF THE STOMACH	56
,, 11.	,, ,,	ROOF OF THE STOMACH, FROM WITHIN	60
,, 12.	,, ,,	DISSECTION FROM THE SIDE (MALE).	62
,, 13.	,, ,,	ALIMENTARY CANAL FROM ABOVE .	65
,, 14.	,, ,,	BLOOD CORPUSCLES	68
,, 15.	,, ,,	TRANSVERSE SECTION OF THORAX ..	70
,, 16.	,, ,,	THE HEART . . .	72
,, 17.	,, ,,	STRUCTURE OF THE GILLS . .	76
,, 18.	,, ,,	THE GREEN GLAND	83

			PAGE
FIG. 19.	*Astacus fluviatilis.*	MUSCULAR TISSUE	91
,, 20.	,, ,,	MUSCLES OF CHELA	93
,, 21.	,, ,,	ARTICULATION OF ABDOMINAL SOMITES	97
,, 22.	,, ,,	MUSCULAR SYSTEM	100
,, 23.	,, ,,	NERVE FIBRES	102
,, 24.	,, ,,	NERVE GANGLIA	103
,, 25.	,, ,,	NERVOUS SYSTEM	104
,, 26.	,, ,,	OLFACTORY AND AUDITORY ORGANS	114
,, 27.	,, ,,	AUDITORY SAC	117
,, 28.	,, ,,	STRUCTURE OF EYE	119
,, 29.	,, ,,	DIAGRAM OF EYE	123
,, 30.	,, ,,	FEMALE REPRODUCTIVE ORGANS	129
,, 31.	,, ,,	MALE REPRODUCTIVE ORGANS	130
,, 32.	,, ,,	STRUCTURE OF OVARY	131
,, 33.	,, ,,	STRUCTURE OF TESTIS	132
,, 34.	,, ,,	SPERMATOZOA	134
,, 35.	,, ,,	THE LAST THORACIC STERNUM IN THE MALE AND FEMALE	136
,, 36.	,, ,,	TRANSVERSE SECTION OF ABDOMEN	142
,, 37.	,, ,,	ABDOMINAL APPENDAGES	144
,, 38.	,, ,,	CONNECTION BETWEEN THORAX AND ABDOMEN	151
,, 39.	,, ,,	CEPHALOTHORACIC STERNA AND ENDOPHRAGMAL SYSTEM	153
,, 40.	,, ,,	OPHTHALMIC AND ANTENNULARY SOMITES	156
,, 41.	,, ,,	THE ROSTRUM	157
,, 42.	,, ,,	A SEGMENT OF THE ENDOPHRAGMAL SYSTEM	159
,, 43.	,, ,,	LONGITUDINAL SECTION OF CEPHALOTHORAX	162
,, 44.	,, ,,	THE THIRD MAXILLIPEDE	164

			PAGE
FIG. 45.	*Astacus fluviatilis.*	THE FIRST AND SECOND MAXILLIPEDES	166
,, 46.	,, ,,	THE SECOND AMBULATORY LEG	169
,, 47.	,, ,,	THE MANDIBLE AND MAXILLÆ	171
,, 48.	,, ,,	THE EYE-STALK, ANTENNULE, AND ANTENNA	172
,, 49.	,, ,,	BLOOD CORPUSCLES	176
,, 50.	,, ,,	EPITHELIUM	178
,, 51.	,, ,,	CONNECTIVE TISSUE	179
,, 52.	,, ,,	MUSCULAR TISSUE	181
,, 53.	,, ,,	MUSCULAR TISSUE	182
,, 54.	,, ,,	NERVE GANGLIA	188
,, 55.	,, ,,	NERVE FIBRES	189
,, 56.	,, ,,	CUTICULAR TISSUE	191
,, 57.	,, ,,	SECTIONS OF EMBRYOS	208
,, 58.	,, ,,	EARLIER STAGES OF DEVELOPMENT	210
,, 59.	,, ,,	LATER STAGES OF DEVELOPMENT	216
,, 60.	,, ,,	NEWLY HATCHED YOUNG	220
,, 61.	,, { *torrentium.* *nobilis.* *nigrescens.* }	COMPARATIVE VIEWS OF THE CARAPACE, THIRD ABDOMINAL SOMITE, AND TELSON	233
,, 62.	,, { *torrentium.* *nobilis.* *nigrescens.* }	COMPARATIVE VIEWS OF THE FIRST AND SECOND ABDOMINAL APPENDAGES OF THE MALE	245
,, 63.	*Cambarus Clarkii*		248
,, 64.	*Parastacus brasiliensis*		250
,, 65.	*Astacoides madagascarensis*		251
,, 66.	DIAGRAM OF THE MORPHOLOGICAL RELATIONS OF THE *Astacidæ*		253
,, 67.	*Homarus vulgaris*		258
,, 68.	{ *Parastacus.* *Nephrops.* *Palæmon.* }	PODOBRANCHIÆ	259

PAGE
FIG. 69. *Nephrops norvegicus* 260
,, 70. *Palinurus vulgaris* 262
,, 71. *Palæmon jamaicensis* 269
,, 72. *Cancer pagurus* 273
,, 73. *Penæus* 281
,, 74. *Cancer pagurus.* DEVELOPMENT 282
,, 75. *Astacus leptodactylis* 301
,, 76. *Australian Crayfish* 307
,, 77. MAP OF THE DISTRIBUTION OF CRAYFISHES . . . 309
,, 78. *Cambarus.* WALKING LEG. 312
,, 79. *Palæmon jamaicensis* 329
,, 80. { *Pseudastacus pustulosus* / *Eryma modestiformis* } 340
,, 81. *Hoploparia longimana* 342

THE CRAYFISH:

AN INTRODUCTION TO THE STUDY OF ZOOLOGY.

CHAPTER I.

THE NATURAL HISTORY OF THE COMMON CRAYFISH

(*Astacus fluviatilis.*)

MANY persons seem to believe that what is termed Science is of a widely different nature from ordinary knowledge, and that the methods by which scientific truths are ascertained involve mental operations of a recondite and mysterious nature, comprehensible only by the initiated, and as distinct in their character as in their subject matter, from the processes by which we discriminate between fact and fancy in ordinary life.

But any one who looks into the matter attentively will soon perceive that there is no solid foundation for the belief that the realm of science is thus shut off from that of common sense; or that the mode of investigation which yields such wonderful results to the scientific investigator, is different in kind from that which is employed for the commonest purposes of everyday existence. Common sense is science exactly in so far as it fulfils the ideal of common sense; that is, sees facts as they are, or, at any rate, without the distortion of prejudice, and reasons from them in accordance with the dictates of sound judgment. And science is simply common sense at its best; that is, rigidly accurate in observation, and merciless to fallacy in logic.

Whoso will question the validity of the conclusions of sound science, must be prepared to carry his scepticism a long way; for it may be safely affirmed, that there is hardly any of those decisions of common sense on which men stake their all in practical life, which can justify itself so thoroughly on common sense principles, as the broad truths of science can be justified.

The conclusion drawn from due consideration of the nature of the case is verified by historical inquiry; and the historian of every science traces back its roots to the primary stock of common information possessed by all mankind.

In its earliest development knowledge is self-sown. Impressions force themselves upon men's senses whether they will or not, and often against their will. The amount of interest which these impressions awaken is determined by the coarser pains and pleasures which they carry in their train, or by mere curiosity; and reason deals with the materials supplied to it as far as that interest carries it, and no farther. Such common knowledge is rather brought than sought; and such ratiocination is little more than the working of a blind intellectual instinct.

It is only when the mind passes beyond this condition that it begins to evolve science. When simple curiosity passes into the love of knowledge as such, and the gratification of the æsthetic sense of the beauty of completeness and accuracy seems more desirable than the easy indolence of ignorance; when the finding out of the causes of things becomes a source of joy, and he is counted happy who is successful in the search; common knowledge of nature passes into what our forefathers called Natural History, from whence there is but a step to that which used to be termed Natural Philosophy, and now passes by the name of Physical Science.

In this final stage of knowledge, the phenomena of nature are regarded as one continuous series of causes and effects; and the ultimate object of science is to trace out that series, from the term which is nearest to us, to that which is at the furthest limit accessible to our means of investigation.

The course of nature as it is, as it has been, and as it will be, is the object of scientific inquiry; whatever lies beyond, above, or below this, is outside science. But the philosopher need not despair at the limitation of his field of labour: in relation to the human mind Nature is boundless; and, though nowhere inaccessible, she is everywhere unfathomable.

The Biological Sciences embody the great multitude of truths which have been ascertained respecting living beings; and as there are two chief kinds of living things, animals and plants, so Biology is, for convenience sake, divided into two main branches, Zoology and Botany.

Each of these branches of Biology has passed through the three stages of development, which are common to all the sciences; and, at the present time, each is in these different stages in different minds. Every country boy possesses more or less information respecting the plants and animals which come under his notice, in the stage of common knowledge; a good many persons have acquired more or less of that accurate, but necessarily incomplete and unmethodised knowledge, which is understood by Natural History; while a few have reached the purely scientific stage, and, as Zoologists and Botanists, strive towards the perfection of Biology as a branch of Physical Science.

Historically, common knowledge is represented by the allusions to animals and plants in ancient literature; while Natural History, more or less grading into Biology, meets us in the works of Aristotle, and his continuators in the Middle Ages, Rondoletius, Aldrovandus, and their contemporaries and successors. But the conscious attempt to construct a complete science of Biology hardly dates further back than Treviranus and Lamarck, at the beginning of this century, while it has received its strongest impulse, in our own day, from Darwin.

My purpose, in the present work, is to exemplify the general truths respecting the development of zoological science which have just been stated by the study of a special case; and, to this end, I have selected an animal, the Common Crayfish, which, taking it altogether, is better fitted for my purpose than any other.

It is readily obtained,* and all the most important points of its construction are easily deciphered; hence, those who read what follows will have no difficulty in ascertaining whether the statements correspond with facts or not. And unless my readers are prepared to take this much trouble, they may almost as well shut the book; for nothing is truer than Harvey's dictum, that those who read without acquiring distinct images of the things about which they read, by the help of their own senses, gather no real knowledge, but conceive mere phantoms and idola.

It is a matter of common information that a number of our streams and rivulets harbour small animals, rarely more than three or four inches long, which are very similar to little lobsters, except that they are usually of a dull, greenish or brownish colour, generally diversified with pale yellow on the under side of the body, and sometimes with red on the limbs. In rare cases, their

* If crayfish are not to be had, a lobster will be found to answer to the description of the former, in almost all points; but the gills and the abdominal appendages present differences; and the last thoracic somite is united with the rest in the lobster. (*See* Chap. V.)

general hue may be red or blue. These are "crayfishes," and they cannot possibly be mistaken for any other inhabitants of our fresh waters.

FIG. 1.—*Astacus fluviatilis.*—Side view of a male specimen (nat. size):—*bg*, branchiostegite; *cg*, cervical groove; *r*, rostrum; *t*, telson.—1, eye-stalk; 2, antennule; 3, antenna; 9, external maxillipede; 10, forceps; 14, last ambulatory leg; 17, third abdominal appendage; 20, lateral lobe of the tail-fin, or sixth abdominal appendage; XV, the first; and XX, the last abdominal somite. In this and in succeeding figures the numbers of the somites are given in Roman, those of the appendages in ordinary numerals.

The animals may be seen walking along the bottom of the shallow waters which they prefer, by means of four pairs of jointed legs (fig. 1); but, if alarmed, they swim backwards with rapid jerks, propelled by the strokes of a broad, fan-shaped flipper, which terminates the hinder end of the body (fig. 1, *t*, *20*). In front of the four pairs of legs, which are used in walking, there is a pair of limbs of a much more massive character, each of which ends in two claws disposed in such a manner as to constitute a powerful pincer (fig. 1; *10*). These claws are the chief weapons of offence and defence of the crayfish, and those who handle them incautiously will discover that their grip is by no means to be despised, and indicates a good deal of disposable energy. A sort of shield covers the front part of the body, and ends in a sharp projecting spine in the middle line (*r*). On each side of this is an eye, mounted on a movable stalk (*1*), which can be turned in any direction: behind the eyes follow two pairs of feelers; in one of these, the feeler ends in two, short, jointed filaments (*2*); while, in the other, it terminates in a single, many-jointed filament, like a whip-lash, which is more than half the length of the body (*3*). Sometimes turned backwards, sometimes sweeping forwards, these long feelers continually explore a considerable area around the body of the crayfish.

If a number of crayfishes, of about the same size, are compared together, it will easily be seen that they fall into two sets; the jointed tail being much broader, especially in the middle, in the one set than in the other (fig. 2). The broad-tailed crayfishes are the females, the others the males. And the latter may be still more easily known by the possession of four curved styles, attached to the under face of the first two rings of the tail, which are turned forwards between the hinder legs, on the under side of the body (fig. 3, A; *15, 16*). In the female, there are mere soft filaments in the place of the first pair of styles (fig. 3, B; *15*).

Crayfishes do not inhabit every British river, and even where they are known to abound, it is not easy to find them at all times of the year. In granite districts and others, in which the soil yields little or no calcareous matter to the waters which flow over it, crayfishes do not occur. They are intolerant of great heat and of much sunshine; they are therefore most active towards the evening, while they shelter themselves under the shade of stones and banks during the day. It has been observed that they frequent those parts of a river which run north and south, less than those which have an easterly and westerly direction, inasmuch as the latter yield more shade from the mid-day sun.

During the depth of winter, crayfishes are rarely to be seen about in a stream; but they may be found in abundance in its banks, in natural crevices and in burrowswhich they dig for themselves. The burrows may be from a few inches to more than a yard deep, and it has been noticed that, if the waters are liable to freeze, the burrows are deeper and further from the surface than otherwise. Where the soil, through which a stream haunted by crayfishes runs, is soft and peaty, the crayfishes work their way into it in all directions, and thousands of them, of all sizes, may be dug out, even at a considerable distance from the banks.

It does not appear that crayfishes fall into a state of torpor in the winter, and thus "hybernate" in the strict sense of the word. At any rate, so long as the weather is open, the crayfish lies at the mouth of his burrow, barring the entrance with his great claws, and with protruded feelers keeps careful watch on the passers-by. Larvæ of insects, water-snails, tadpoles, or frogs, which come within reach, are suddenly seized and devoured, and it is averred that the water-rat is liable to the same fate. Passing too near the fatal den, possibly in search of a stray crayfish, whose flavour he highly appreciates, the vole is himself seized and held till he is suffocated, when his captor easily reverses the conditions of the anticipated meal.

In fact, few things in the way of food are amiss to the crayfish; living or dead, fresh or carrion, animal or vegetable, it is all one. Calcareous plants, such as the stoneworts (*Chara*), are highly acceptable; so are any kinds of succulent roots, such as carrots; and it is said that crayfish sometimes make short excursions inland, in search of vegetable food. Snails are devoured, shells and all; the cast coats of other crayfish are turned to account as supplies of needful calcareous matter; and the unprotected or weakly member of the family is

2

not spared. Crayfishes, in fact, are guilty of cannibalism in its worst form; and a French observer pathetically remarks, that, under certain circumstances, the males "*méconnaissent les plus saints devoirs;*" and, not content with mutilating or killing their spouses, after the fashion of animals of higher moral pretensions, they descend to the lowest depths of utilitarian turpitude, and finish by eating them.

In the depth of winter, however, the most alert of crayfish can find little enough food; and hence, when they emerge from their hiding-places in the first warm days of spring, usually about March, the crayfishes are in poor condition.

At this time, the females are found to be laden with eggs, of which from one to two hundred are attached beneath the tail, and look like a mass of minute berries (fig. 3, B). In May or June, these eggs are hatched, and give rise to minute young, which are sometimes to be found attached beneath the tail of the mother, under whose protection they spend the first few days of their existence.

In this country, we do not set much store upon crayfishes as an article of food, but on the Continent, and especially in France, they are in great request. Paris alone, with its two millions of inhabitants, consumes annually from five to six millions of crayfishes, and pays about £16,000 for them. The natural productivity of the rivers of France has long been inadequate to supply the demand for these delicacies; and hence, not only are large quantities imported from Germany, and elsewhere, but the artificial cultivation of crayfish has been successfully attempted on a considerable scale.

Crayfishes are caught in various ways; sometimes the fisherman simply wades in the water and drags them out of their burrows; more commonly, hoop-nets baited with frogs are let down into the water and rapidly drawn up, when there is reason to think that crayfish have been attracted to the bait; or fires are lighted on the banks at night, and the crayfish, which are attracted, like moths, to the unwonted illumination, are scooped out with the hand or with nets.

Thus far, our information respecting the crayfish is such as would be forced upon anyone who dealt in crayfishes, or lived in a district in which they were commonly used for food. It is common knowledge. Let us now try to push our acquaintance with what is to be learned about the animal a little further, so as to be able to give an account of its Natural History, such as might have been furnished by Buffon if he had dealt with the subject.

There is an inquiry which does not strictly lie within the province of physical science, and yet suggests itself naturally enough at the outset of a natural history.

The animal we are considering has two names, one common, *Crayfish,* the other technical, *Astacus fluviatilis.* How has it come by these two names, and why, having a common English name for it already, should naturalists call it by another appellation derived from a foreign tongue?

The origin of the common name, "crayfish," involves some curious questions of etymology, and indeed, of history. It might readily be supposed that the word "cray" had a meaning of its own, and qualified the substantive "fish"—as "jelly" and "cod" in "jellyfish" and "codfish." But this certainly is not the case. The old English method of writing the word was "crevis" or "crevice," and the "cray" is simply a phonetic spelling of the syllable "cre," in which the "e" was formerly pronounced as all the world, except ourselves, now pronounce that vowel. While "fish" is the "vis" insensibly modified to suit our knowledge of the thing as an aquatic animal.

Now "crevis" is clearly one of two things. Either it is a modification of the French name "écrevisse," or of the Low Dutch name "crevik," by which the crayfish is known in these languages. The former derivation is that usually given, and, if it be correct, we must refer "crayfish" to the same category as "mutton," "beef," and "pork," all of which are French equivalents, introduced by the Normans, for the "sheep's flesh," "ox flesh," and "swine's flesh," of their English subjects. In this case, we should not have called a crayfish, a crayfish, except for the Norman conquest.

On the other hand, if "crevik" is the source of our

word, it may have come to us straight from the Angle and Saxon contingent of our mixed ancestry.

As to the origin of the technical name; ἀστακός, *astakos*, was the name by which the Greeks knew the lobster; and it has been handed down to us in the works of Aristotle, who does not seem to have taken any special notice of the crayfish. At the revival of learning, the early naturalists noted the close general similarity between the lobster and the crayfish; but, as the latter lives in fresh water, while the former is a marine animal, they called the crayfish, in their Latin, *Astacus fluviatilis*, or the "river-lobster," by way of distinction; and this nomenclature was retained until, about forty-five years ago, an eminent French Naturalist, M. Milne-Edwards, pointed out that there are far more extensive differences between lobsters and crayfish than had been supposed; and that it would be advisable to mark the distinctness of the things by a corresponding difference in their names. Leaving *Astacus* for the crayfishes, he proposed to change the technical name of the lobster into *Homarus*, by latinising the old French name "*Omar*," or "*Homar*" (now *Homard*), for that animal.

At the present time, therefore, while the recognised technical name of the crayfish is *Astacus fluviatilis*, that of the lobster is *Homarus vulgaris*. And as this nomenclature is generally received, it is desirable that it should not be altered; though it is attended by the inconvenience, that *Astacus*, as we now employ the name, does not denote that which the Greeks, ancient and modern, signify, by its original, *astakos;* and does signify something quite different.

Finally, as to why it is needful to have two names for the same thing, one vernacular, and one technical. Many people imagine that scientific terminology is a needless burden imposed upon the novice, and ask us why we cannot be content with plain English. In reply, I would suggest to such an objector to open a conversation about his own business with a carpenter, or an engineer, or, still better, with a sailor, and try how far plain English will go. The interview will not have lasted long before he will find himself lost in a maze of unintelligible technicalities. Every calling has its technical terminology; and every artisan uses terms of art, which sound like gibberish to those who know nothing of the art, but are exceedingly convenient to those who practise it.

In fact, every art is full of conceptions which are special to itself; and, as the use of language is to convey our conceptions to one another, language must supply signs for those conceptions. There are two ways of doing this: either existing signs may be combined in loose and cumbrous periphrases; or new signs, having a well-understood and definite signification, may be invented. The practice of sensible people shows the advantage of the latter course; and here, as elsewhere, science has simply followed and improved upon common sense.

Moreover, while English, French, German, and Italian artisans are under no particular necessity to discuss the processes and results of their business with one another, science is cosmopolitan, and the difficulties of the study of Zoology would be prodigiously increased, if Zoologists of different nationalities used different technical terms for the same thing. They need a universal language; and it has been found convenient that the language shall be the Latin in form, and Latin or Greek in origin. What in English is Crayfish, is *Écrevisse* in French; *Flusskrebs*, in German; *Cammaro*, or *Gambaro*, or *Gammarello*, in Italian: but the Zoologist of each nationality knows that, in the scientific works of all the rest, he shall find what he wants to read under the head of *Astacus fluviatilis*.

But granting the expediency of a technical name for the Crayfish, why should that name be double? The reply is still, practical convenience. If there are ten children of one family, we do not call them all Smith, because such a procedure would not help us to distinguish one from the other; nor do we call them simply John, James, Peter, William, and so on, for that would not help us to identify them as of one family. So we give them all two names, one indicating their close relation, and the other their separate individuality —as John Smith, James Smith, Peter Smith, William Smith, &c. The same thing is done in Zoology; only, in accordance with the genius of the Latin language, we put the Christian name, so to speak, after the surname.

There are a number of kinds of Crayfish, so similar to one another that they bear the common surname of *Astacus*. One kind, by way of distinction, is called *fluviatile*, another *slender-handed*, another *Dauric*, from the region in which it lives; and these double names are rendered by—*Astacus fluviatilis*, *Astacus leptodactylus*, and *Astacus dauricus;* and thus we have a nomenclature which is exceedingly simple in principle, and free from confusion in practice. And I may add that, the less attention is paid to the original meaning of the substantive and adjective terms of this binomial nomenclature, and the sooner they are used as proper names, the better. Very good reasons for using a term may exist when it is first invented, which lose their validity with the progress of knowledge. Thus *Astacus fluviatilis* was a significant name so long as we knew of only one kind of crayfish; but now that we are acquainted with a number of kinds, all of which inhabit rivers, it is meaningless. Nevertheless, as changing it would involve endless confusion, and the object of nomenclature is simply to have a definite name for a definite thing, nobody dreams of proposing to alter it.

Having learned this much about the origin of the names of the crayfish, we may next proceed to consider those points which an observant Naturalist, who did not care to go far beyond the surface of things, would find to notice in the animal itself.

Probably the most conspicuous peculiarity of the crayfish, to any one who is familiar only with the higher animals, is the fact that the hard parts of the body are outside and the soft parts inside; whereas in ourselves, and in the ordinary domestic animals, the hard parts, or bones, which constitute the skeleton, are inside, and the soft parts clothe them. Hence, while our hard framework is said to be an *endoskeleton*, or internal skeleton; that of the crayfish is termed an *exoskeleton*, or external skeleton. It is from the circumstance that the body of the crayfishes is enveloped in this hard crust, that the name of *Crustacea* is applied to them, along with the crabs, shrimps, and other such animals. Insects, spiders, and centipedes have also a hard exoskeleton, but it is usually not so hard and thick as in the *Crustacea*.

If a piece of the crayfish's skeleton is placed in strong vinegar, abundant bubbles of carbonic acid gas are given off from it, and it rapidly becomes converted into a soft laminated membrane, while the solution will be found to contain lime. In fact the exoskeleton is composed of a peculiar animal matter, so much impregnated with carbonate and phosphate of lime that it becomes dense and hard.

It will be observed that the body of the crayfish is naturally marked out into several distinct regions. There

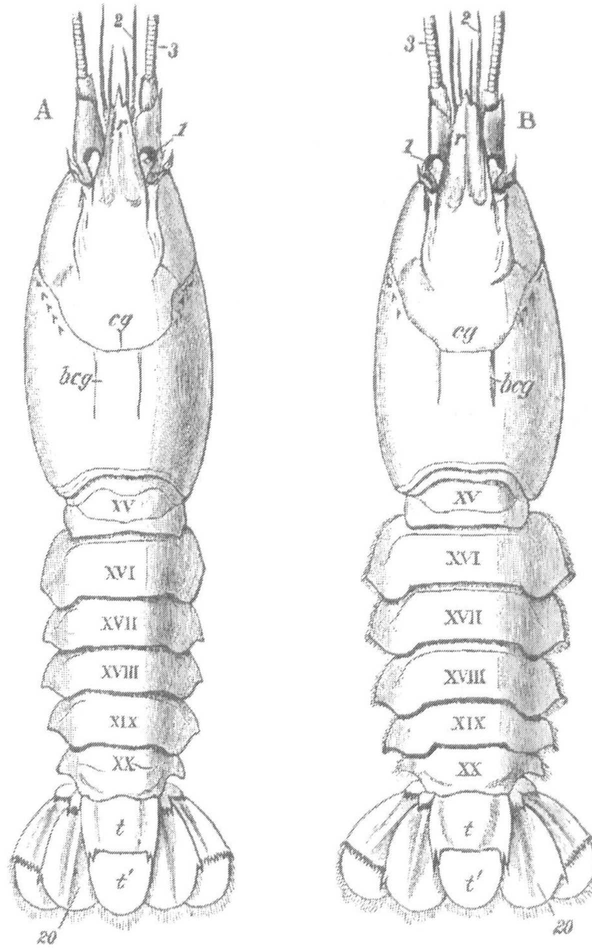

FIG. 2.—*Astacus fluviatilis.*—Dorsal or tergal views (nat. size). A, male; B, female:—*bcg*, branchio-cardiac groove, which marks the boundary between the pericardial and the branchial cavities; *cg*, cervical groove; these letters are placed on the carapace; *r*, rostrum; *t*, *t′*, the two divisions of the telson; *1*, eye-stalks; *2*, antennules; *3*, antennæ; *20*, lateral lobes of tail-fin; XV–XX, somites of the abdomen.

is a firm and solid front part, covered by a large continuous shield, which is called the *carapace;* and a jointed hind part, commonly termed the tail (fig. 2). From the perception of a partially real, and partially fanciful, analogy with the regions into which the body is divided in the higher animals, the fore part is termed the *cephalo-thorax,* or head (*cephalon*) and chest (*thorax*) combined, while the hinder part receives the name of *abdomen.*

Now the exoskeleton is not of the same constitution throughout these regions. The abdomen, for example, is composed of six complete hard rings (fig. 2, xv–xx), and a terminal flap, on the under side of which the vent (fig. 3, *a*) is situated, and which is called the *telson* (fig. 2, *t*, *t'*). All these are freely moveable upon one another, inasmuch as the exoskeleton which connects them is not calcified, but is, for the most part, soft and flexible, like the hard exoskeleton when the lime salts have been removed by acid. The mechanism of the joints will have to be attentively considered by-and-by; it is sufficient, at present, to remark that, wherever a joint exists, it is produced in the same fashion, by the exoskeleton remaining soft in certain regions of the jointed part.

The carapace is not jointed; but a transverse groove is observed about the middle of it, the ends of which run down on the sides and then turn forwards (figs. 1 and 2, *cg*). This is called the *cervical groove,* and it marks off the region of the head, in front, from that of the thorax behind.

The thorax seems at first not to be jointed at all; but if its under, or what is better called its *sternal*, surface is examined carefully, it will be found to be divided into as many transverse bands, or segments, as there are pairs of legs (fig. 3); and, moreover, the hindermost of these segments is not firmly united with the rest, but can be moved backwards and forwards through a small space (fig. 3, B; xiv).

Attached to the sternal side of every ring of the abdomen of the female there is a pair of limbs, called *swimmerets*. In the five anterior rings, these are small and slender (fig. 3, B; *15, 19*); but those of the sixth ring are very large, and each ends in two broad plates (*20*). These two plates on each side, with the telson in the middle, constitute the flapper of the crayfish, by the aid of which it executes its retrograde swimming movements. The small swimmerets move together with a regular swing, like paddles, and probably aid in propelling the animal forwards. In the breeding female (B), the eggs are attached to them; while, in the male, the two anterior pairs (A; *15, 16*) are converted into the peculiar styles which distinguish that sex.

The four pairs of legs which are employed for walking purposes, are divided into a number of joints, and the foremost two pairs are terminated by double claws, arranged so as to form a pincer, whence they are said to

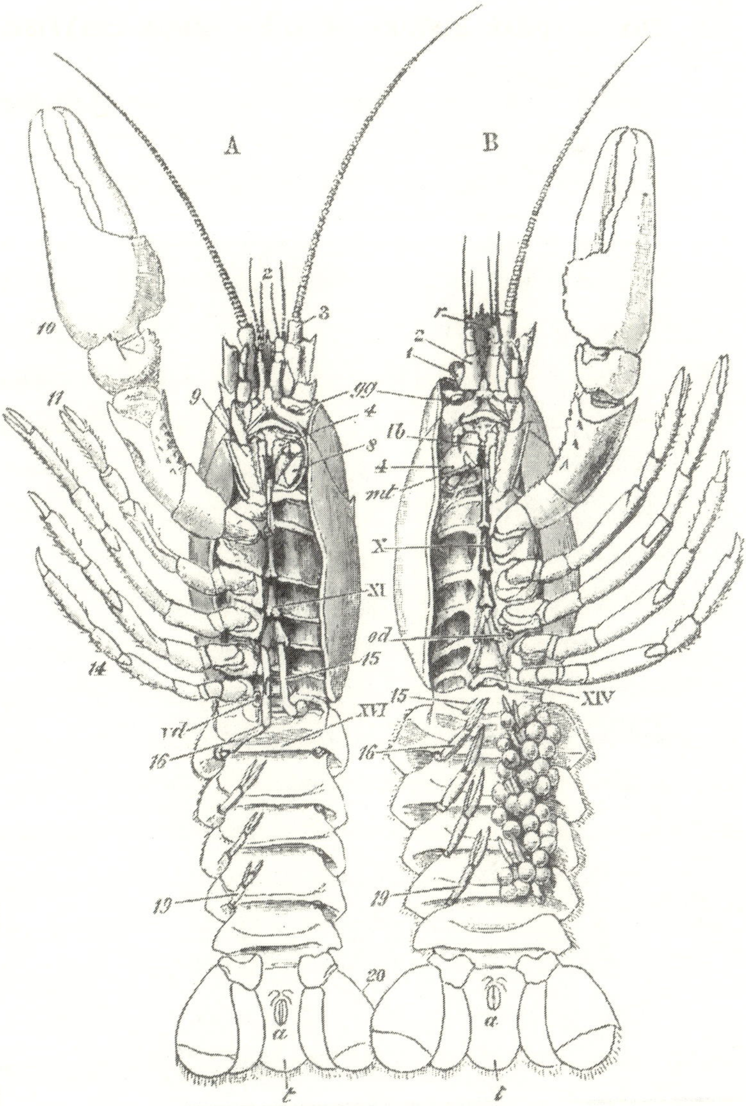

FIG. 3.—*Astacus fluviatilis.*—Ventral or sternal views (nat. size). A, male; B, female:—*a*, vent; *gg*, opening of the green gland; *lb*, labrum; *mt*, metastoma or lower lip; *od*, opening of the oviduct; *vd*, that of the vas deferens. *1*, eye-stalk; *2*, antennule; *3*, antenna; *4*, mandible; *8*, second maxillipede; *9*, third or external maxillipede; *10*, forceps; *11*, first leg; *14*, fourth leg; *15*, *16*, *19*, *20*, first, second, fifth, and sixth abdominal appendages; X., XI., XIV., sterna of the fourth, fifth, and eighth thoracic somite; XVI., sternum of the second abdominal somite. In the male, the 9th to the 14th and the 16th to the 19th appendages are removed on the animal's left side: in the female, the antenna (with the exception of its basal joint) and the 5th to the 14th appendages on the animal's right are removed; the eggs also are shown attached to the swimmerets of the left side of the body.

be *chelate*. The two hindermost pairs, on the other hand, end in simple claws.

In front of these legs, come the great prehensile limbs (*10*), which are chelate, like those which immediately follow them, but vastly larger. They often receive the special name of *chelæ;* and the large terminal joints are called the "hand." We shall escape confusion if we call these limbs the *forceps*, and restrict the name of *chela* to the two terminal joints.

All the limbs hitherto mentioned subserve locomotion and prehension in various degrees. The crayfish swims by the help of its abdomen, and the hinder pairs of abdominal limbs; walks by means of the four hinder pairs of thoracic limbs; lays hold of anything to fix itself, or to assist in climbing, by the two chelate anterior pairs of these limbs, which are also employed in tearing the food seized by the forceps and conveying it to the mouth; while it seizes its prey and defends itself with the forceps. The part which each of these limbs plays is termed its *function*, and it is said to be the *organ* of that function; so that all these limbs may be said to be organs of the functions of locomotion, of offence and defence.

In front of the forceps, there is a pair of limbs which have a different character, and take a different direction from any of the foregoing (*9*). These limbs, in fact, are turned directly forwards, parallel with one another, and with the middle line of the body. They are divided into a number of joints, of which one of those near the base is longer than the rest, and strongly toothed along the inner edge, or that which is turned towards its fellow. It is obvious that these two limbs are well adapted to crush and tear whatever comes between them, and they are, in fact, *jaws* or organs of manducation. At the same time, it will be noticed that they retain a curiously close general resemblance to the hinder thoracic legs; and hence, for distinction's sake, they are called outer *foot-jaws*, or external *maxillipedes*.

If the head of a stout pin is pushed between these external maxillipedes, it will be found that it passes without any difficulty into the interior of the body, through the mouth. In fact, the mouth is relatively rather a large aperture; but it cannot be seen without forcing aside, not only these external foot-jaws, but a number of other limbs, which subserve the same function of manducation, or chewing and crushing the food. We may pass by the organs of manducation, for the present, with the remark that there are altogether three pairs of maxillipedes, followed by two pairs of somewhat differently formed *maxillæ*, and one pair of very stout and strong jaws, which are termed the *mandibles* (4). All these jaws work from side to side, in contradistinction to the jaws of vertebrated animals, which move up and down. In front of, and above the mouth, with the jaws which cover it, are seen the long feelers, which are called the *antennæ* (3); above, and in front of them, follow the small feelers, or *antennules* (2); and over them, again, lie the *eye stalks* (*1*). The antennæ are organs of touch; the antennules, in addition, contain the organs of hearing; while, at the ends of the eyestalks, are the organs of vision.

Thus we see that the crayfish has a jointed and segmented body, the rings of which it is composed being very obvious in the abdomen, but more obscurely traceable elsewhere; that it has no fewer than twenty pairs of what may be called by the general name of *appendages;* and that these appendages are turned to different uses, or are organs of different functions, in different parts of the body. The crayfish is obviously a very complicated piece of living machinery. But we have not yet come to the end of all the organs that may be discovered even by cursory inspection. Every one who has eaten a boiled crayfish, or a lobster, knows that the great shield, or carapace, is very easily separated from the thorax and abdomen, the head and the limbs which belong to that region coming away with the carapace. The reason of this is not far to seek. The lower edges of that part of the carapace which belongs to the thorax approach the bases of the legs pretty closely, but a cleft-like space is left; and this cleft extends forwards to the sides of the region of the mouth, and backwards and upwards, between the hinder margin of the carapace and the sides of the first ring of the abdomen, which are partly overlapped by, and partly overlap, that margin. If the blade of a pair of scissors is carefully introduced into the cleft from behind, as high up as it will go without tearing anything, and a cut is then made, parallel with the middle line, as far as the cervical groove, and thence following the cervical groove to the base of the outer foot-jaws, a large flap will be removed. This flap of the carapace is called the *branchiostegite* (fig. 1, *bg*), because it covers the gills or *branchiæ* (fig. 4), which are now exposed. They have the appearance of a number of delicate plumes, which take a direction from the bases of the legs upwards and forwards behind, upwards and backwards in front, their summits converging towards the upper end of the cavity in which they are placed, and which is called the *branchial chamber*. These branchiæ are the respiratory organs; and they perform the same functions as the gills of a fish, to which they present some similarity.

If the gills are cleared away, it is seen that the branchial cavity is bounded, on the inner side, by a sloping wall, formed by a delicate, but more or less calcified layer of the exoskeleton, which constitutes the proper outer wall of the thorax. At the upper limit of the branchial cavity, the layer of exoskeleton is very thin, and turning outwards, is continued into the inner wall or lining of the branchiostegite, which is also very thin (*see* fig. 15, p. 70).

Thus the branchial chamber is altogether outside the body, to which it stands in somewhat the same relation as the space between the flaps of a man's coat and his waistcoat would do to the part of the body enclosed by the

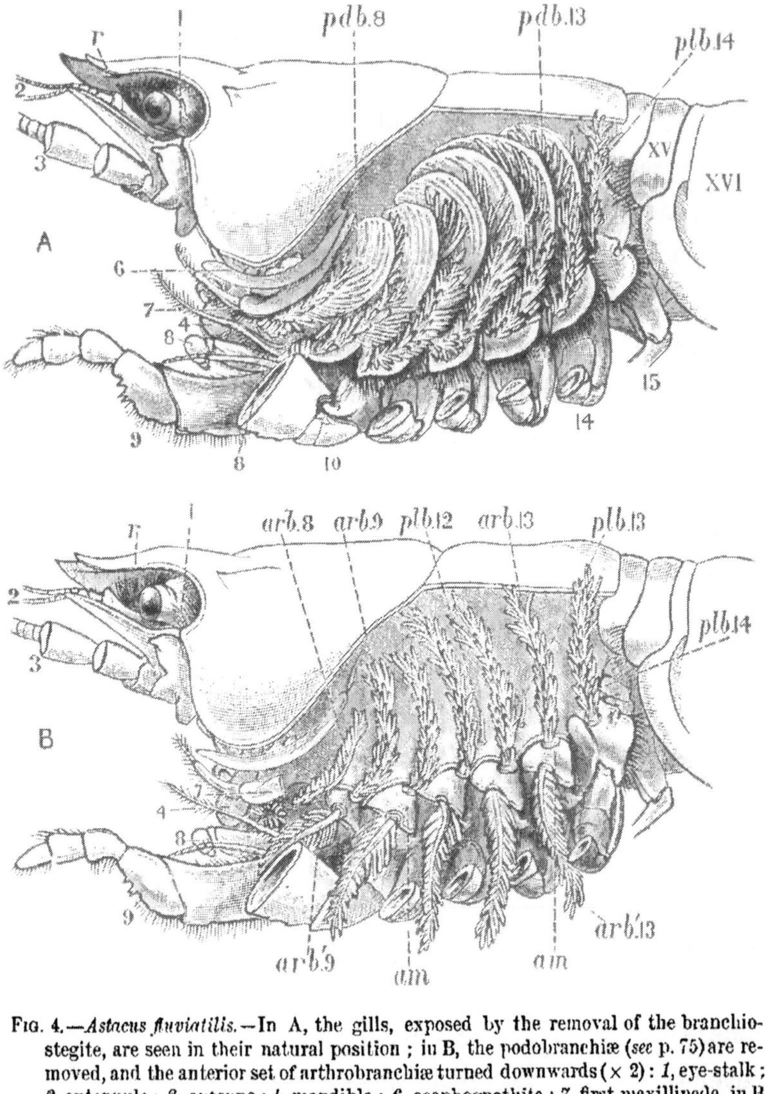

FIG. 4.—*Astacus fluviatilis.*—In A, the gills, exposed by the removal of the branchiostegite, are seen in their natural position; in B, the podobranchiæ (*see* p. 75) are removed, and the anterior set of arthrobranchiæ turned downwards (× 2): *1*, eye-stalk; *2*, antennule; *3*, antenna; *4*, mandible; *6*, scaphognathite; *7*, first maxillipede, in B the epipodite, to which the line points, is partly removed; *8*, second maxillipede; *9*, third maxillipede; *10*, forceps; *14*, fourth ambulatory leg; *15*, first abdominal appendage; XV., first, and XVI., second abdominal somite; *arb. 8*, *arb. 9*, *arb. 13*, the posterior arthrobranchiæ of the second and third maxillipedes and of the third ambulatory leg; *arb'. 9*, *arb'. 13*, the anterior arthrobranchiæ of the third maxillipede and of the third ambulatory leg; *pbd. 8*, podobranchiæ of the second maxillipede; *pbd. 13*, that of the third ambulatory leg; *plb. 12*, *plb. 13*, the two rudimentary pleurobranchiæ; *plb. 14*, the functional pleurobranchia; *r*, rostrum.

waistcoat, if we suppose the lining of the flaps to be made in one piece with the sides of the waistcoat. Or a closer parallel still would be brought about, if the skin of a man's back were loose enough to be pulled out, on each side, into two broad flaps covering the flanks.

It will be observed that the branchial chamber is open behind, below, and in front; and, therefore, that the water in which the crayfish habitually lives has free ingress and egress. Thus the air dissolved in the water enables breathing to go on, just as it does in fishes. As is the case with many fishes, the crayfish breathes very well out of the water, if kept in a situation sufficiently cool and moist to prevent the gills from drying up; and thus there is no reason why, in cool and damp weather, the crayfish should not be able to live very well on land, at any rate among moist herbage, though whether our common crayfishes do make such terrestrial excursions is perhaps doubtful. We shall see, by-and-by, that there are some exotic crayfish which habitually live on land, and perish if they are long submerged in water.

With respect to the internal structure of the crayfish, there are some points which cannot escape notice, however rough the process of examination may be.

Thus, when the carapace is removed in a crayfish which has been just killed, the heart is seen still pulsating. It is an organ of considerable relative size (fig. 5, *h*), which is situated immediately beneath the

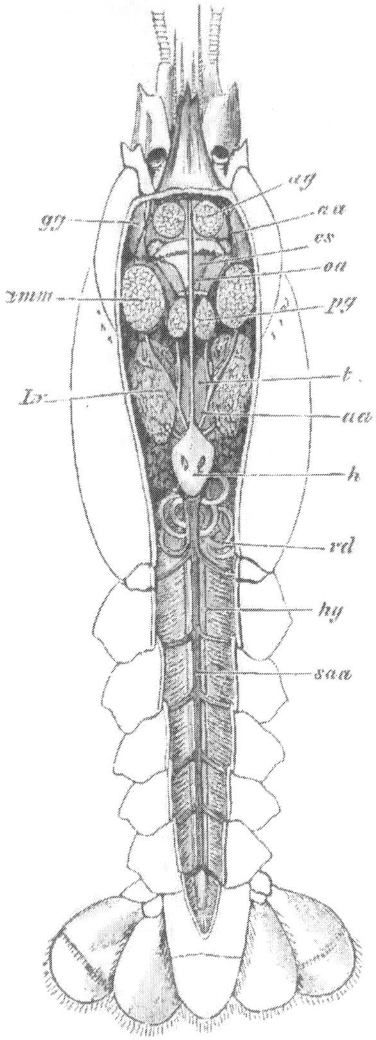

FIG. 5.—*Astacus fluviatilis.*—A male specimen, with the roof of the carapace and the terga of the abdominal somites removed to show the viscera (nat. size) :—*aa*, antennary artery ; *ag*, anterior gastric muscles ; *amm*, adductor muscles of the mandibles; *cs*, cardiac portion of the stomach ; *gg*, green glands ; *h*, heart ; *hg*, hind gut, or large intestine ; *Lr*, liver ; *oa*, ophthalmic artery ; *pg* posterior gastric muscles ; *saa*, superior abdominal artery ; *t*, testis ; *vd*, vas deferens.

middle region of that part of the carapace which lies behind the cervical groove; or, in other words, in the dorsal region of the thorax. In front of it, and therefore in the head, is a large rounded sac, the stomach (fig. 5, *cs*; fig. 6, *cs*, *ps*), from which a very delicate intestine (figs. 5 and 6, *hg*) passes straight back through the thorax and abdomen to the vent (fig. 6, *a*).

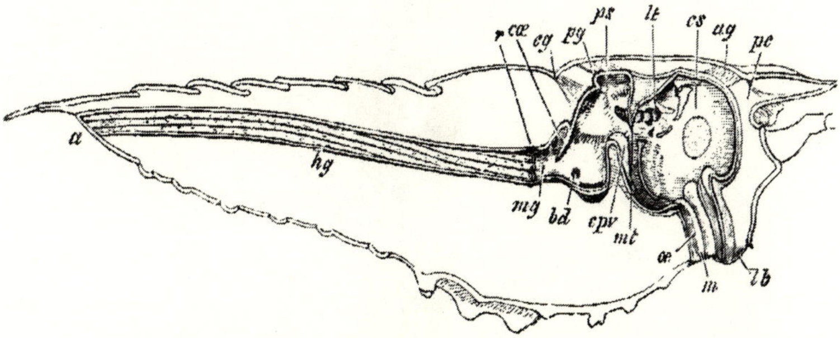

FIG. 6.—*Astacus fluviatilis.*—A longitudinal vertical section of the alimentary canal, with the outline of the body (nat. size):—*a*, vent; *ag*, anterior gastric muscle; *bd*, entrance of left bile duct; *cg*, cervical groove; *cæ*, cæcum; *cpv*, cardio-pyloric valve; *cs*, cardiac portion of stomach; the circular area immediately below the end of the line from *cs* marks the position of the gastrolith of the left side; *hg*, hind-gut; *lb*, labrum; *lt*, lateral tooth of stomach; *m*, mouth; *mg*, mid-gut; *mt*, median tooth; *œ*, œsophagus; *pc*, procephalic process; *pg*, posterior gastric muscle; *ps*, pyloric portion of stomach; *r*, annular ridge, marking the commencement of the hind-gut.

In summer, there are commonly to be found at the sides of the stomach two lenticular calcareous masses, which are known as "crabs'-eyes," or *gastroliths*, and were, in old times, valued in medicine as sovereign remedies for all sorts of disorders. These bodies (fig. 7) are smooth and flattened, or concave, on the side which is turned towards the cavity of the stomach; while the opposite side, being convex and rough with irregular prominences, is something like a "brain-stone" coral.

Moreover, when the stomach is laid open, three large

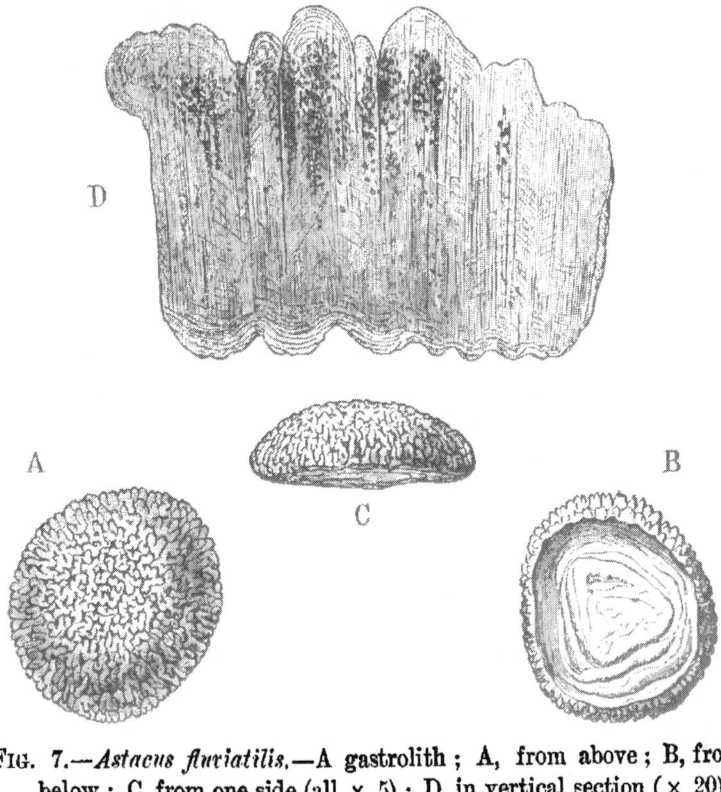

FIG. 7.—*Astacus fluviatilis.*—A gastrolith; A, from above; B, from below; C, from one side (all × 5); D, in vertical section (× 20).

reddish teeth are seen to project conspicuously into its interior (fig. 6, *lt*, *mt*); so that, in addition to its six pairs of jaws, the crayfish has a supplementary crushing mill in its stomach. On each side of the stomach, there is a soft yellow or brown mass, commonly known as the liver (fig. 5, *Lr*); and, in the breeding season, the ovaries of the females, or organs in which the eggs are formed, are very conspicuous from the dark-coloured eggs which they contain, and which, like the exoskeleton, turn red when they are boiled. The corresponding part in a cooked lobster goes by the name of the "coral."

Beside these internal structures, the most noticeable are the large masses of flesh, or muscle, in the thorax and abdomen, and in the pincers; which, instead of being red, as in most of the higher animals, is white. It will further be observed that the blood, which flows readily when a crayfish is wounded, is a clear fluid, and is either almost colourless, or of a very pale reddish or neutral tint. Hence the older Naturalists thought that the crayfish was devoid of blood, and had merely a sort of ichor in place of it. But the fluid in question is true blood; and if it is received into a vessel, it soon forms a soft, but firm, gelatinous clot.

The crayfish grows rapidly in youth, but enlarges more and more slowly as age advances. The young animal which has just left the egg is of a greyish colour, and about one quarter of an inch long. By the end of the year, it may have reached nearly an inch and a half in length. Crayfishes of a year old are, on an average, two inches long; at two years, two inches and four-fifths; at three years, three inches and a half; at four years, four inches and a half nearly; and at five years, five inches. They go on growing till, in exceptional cases, they may attain between seven inches and eight inches in length; but at what degree of longevity this unusual dimension is reached is uncertain. It seems probable, however, that the life of these animals may be prolonged to as much as fifteen or twenty years. They appear to reach maturity, so far as the power of reproduction is concerned, in their fifth or, more usually, their sixth year. However, I have seen a female, with eggs attached under the abdomen, only two inches long, and therefore, probably, in her second year. The males are commonly larger than females of the same age.

The hard skeleton of a crayfish, once formed, is incapable of being stretched, nor can it increase by interstitial addition to its substance, as the bone of one of the higher animals grows. Hence it follows, that the enlargement of the body, which actually takes place, involves the shedding and reproduction of its investment. This might be effected by insensible degrees, and in different parts of the body at different times, as we shed our hair; but, as a matter of fact, it occurs periodically and universally, somewhat as the feathers of birds are moulted. The whole of the old coat of the body is thrown off at once, and suddenly; and the new coat, which has, in the meanwhile, been formed beneath the old one, remains soft for a time, and allows of a rapid increase in the dimensions of the body before it hardens. This sort of moulting is what is technically termed *ecdysis*, or *exuviation*. It is commonly spoken of as the "shedding of the skin," and there is no harm in using this phrase, if we recollect that the shed coat is not the skin, in the proper sense of the word, but only what is termed a *cuticular layer*, which is secreted upon the outer surface of the true integument. The cuticular skeleton of the crayfish, in fact, is not even so much a part of the skin as the cast of a snake, or as our own nails. For these are composed of coherent, formed parts of the epidermis; while the hard investment of the crayfish contains no such formed parts, and is developed on the outside of those structures which answer to the constituents of the epidermis in the higher animals. Thus the crayfish grows, as it were, by starts; its dimensions remaining stationary in the intervals of its moults, and then rapidly increasing for a few days, while the new exoskeleton is in the course of formation.

The ecdysis of the crayfish was first thoroughly studied a century and a half ago, by one of the most accurate observers who ever lived, the famous Réaumur, and the following account of this very curious process is given nearly in his words.*

A few hours before the process of exuviation com-

* See Réaumur's two Memoirs, "Sur les diverses reproductions qui se font dans les écrevisses, les omars, les crabes, etc.," "Histoire de l'Académie royale des Sciences," année 1712; and "Additions aux observations sur la mue des écrevisses données dans les Mémoires de 1712." Ibid. 1718.

mences, the crayfish rubs its limbs one against the other, and, without changing its place, moves each separately, throws itself on its back, bends its tail, and then stretches it out again, at the same time vibrating its antennæ. By these movements, it gives the various parts a little play in their loosened sheaths. After these preparatory steps, the crayfish appears to become distended; in all probability, in consequence of the commencing retraction of the limbs into the interior of the exoskeleton of the body. In fact, it has been remarked, that if, at this period, the extremity of one of the great claws is broken off, it will be found empty, the contained soft parts being retracted as far as the second joint. The soft membranous part of the exoskeleton, which connects the hinder end of the carapace with the first ring of the abdomen, gives way, and the body, covered with the new soft integument, protrudes; its dark brown colour rendering it easily distinguishable from the greenish-brown old integument.

Having got thus far, the crayfish rests for a while, and then the agitation of the limbs and body recommences. The carapace is forced upwards and forwards by the protrusion of the body, and remains attached only in the region of the mouth. The head is next drawn backwards, while the eyes and its other appendages are extracted from their old investment. Next the legs are pulled out, either one at a time, or those of one, or both, sides together. Sometimes a limb gives way and is left behind in its sheath.

The operation is facilitated by the splitting of the old integument of the limb along one side longitudinally.

When the legs are disengaged, the animal draws its head and limbs completely out of their former covering; and, with a sudden spring forward, while it extends its abdomen, it extracts the latter, and leaves its old skeleton behind. The carapace falls back into its ordinary position, and the longitudinal fissures of the sheaths of the limbs close up so accurately, that the shed integument has just the appearance the animal had when the exuviation commenced. The cast exoskeleton is so like the crayfish itself, when the latter is at rest, that, except for the brighter colour of the latter, the two cannot be distinguished.

After exuviation, the owner of the cast skin, exhausted by its violent struggles, which are not unfrequently fatal, lies in a prostrate condition. Instead of being covered by a hard shell, its integument is soft and flabby, like wet paper; though Réaumur remarks, that if a crayfish is handled immediately after exuviation, its body feels hard; and he ascribes this to the violent contraction which its muscles have undergone, leaving them in a state of cramp. In the absence of the hard skeleton, however, there is nothing to bring the contracted muscles at once back into position, and it must be some time before the pressure of the internal fluids is so distributed as to stretch them out.

When the process of exuviation has proceeded so far

that the carapace is raised, nothing stops the crayfish from continuing its struggles. If taken out of the water in this condition, they go on moulting in the hand, and even pressure on their bodies will not arrest their efforts.

The length of time occupied from the first giving way of the integuments to the final emergence of the animal, varies with its vigour, and the conditions under which it is placed, from ten minutes to several hours. The chitinous lining of the stomach, with its teeth, and the "crabs'-eyes," are shed along with the rest of the cuticular exoskeleton; but they are broken up and dissolved in the stomach.

The new integuments of the crayfish remain soft for a period which varies from one to three days; and it is a curious fact, that the animal appears to be quite aware of its helplessness, and governs itself accordingly.

An observant naturalist says: "I once had a domesticated crayfish (*Astacus fluviatilis*), which I kept in a glass pan, in water, not more than an inch and a half deep, previous experiment having shown that in deeper water, probably from want of sufficient aëration, this animal would not live long. By degrees my prisoner became very bold, and when I held my fingers at the edge of the vessel, he assailed them with promptness and energy. About a year after I had him, I perceived, as I thought, a second crayfish with him. On examination, I found it to be his old coat, which he had left in a most perfect state. My friend had now lost his heroism, and fluttered about in the greatest agitation. He was quite soft; and every time I entered the room during the next two days, he exhibited the wildest terror. On the third, he appeared to gain confidence, and ventured to use his nippers, though with some timidity, and he was not yet quite so hard as he had been. In about a week, however, he became bolder than ever; his weapons were sharper, and he appeared stronger, and a nip from him was no joke. He lived in all about two years, during which time his food was a very few worms at very uncertain times; perhaps he did not get fifty altogether."*

It would appear, from the best observations that have yet been made, that the young crayfish exuviate two or three times in the course of the first year; and that, afterwards, the process is annual, and takes place usually about midsummer. There is reason to suppose that very old crayfish do not exuviate every year.

It has been stated that, in the course of its violent efforts to extract its limbs from the cast-off exoskeleton, the crayfish sometimes loses one or other of them; the limb giving way, and the greater part, or the whole, of it remaining in the exuviæ. But it is not only in this way that crayfishes part with their limbs. At all times, if the animal is held by one of its pincers, so that it cannot get away, it is apt to solve the difficulty by casting off

* The late Mr. Robert Ball, of Dublin, in Bell's "British Crustacea," p. 239.

the limb, which remains in the hand of the captor, while the crayfish escapes. This voluntary amputation is always effected at the same place; namely, where the limb is slenderest, just beyond the articulation which unites the basal joint with the next. The other limbs also readily part at the joints; and it is very common to meet with crayfish which have undergone such mutilation. But the injury thus inflicted is not permanent, as these animals possess the power of reproducing lost parts to a marvellous extent, whether the loss has been inflicted by artificial amputation, or voluntarily.

Crayfishes, like all the *Crustacea*, bleed very freely when wounded; and if one of the large joints of a leg is cut through, or if the animal's body is injured, it is very likely to die rapidly from the ensuing hæmorrhage. A crayfish thus wounded, however, commonly throws off the limb at the next articulation, where the cavity of the limb is less patent, and its sides more readily fall together; and, as we have seen, the pincers are usually cast off at their narrowest point. When such amputation has taken place, a crust, probably formed of coagulated blood, rapidly forms over the surface of the stump; and, eventually, it becomes covered with a cuticle. Beneath this, after a time, a sort of bud grows out from the centre of the surface of the stump, and gradually takes on the form of as much of the limb as has been removed. At the next ecdysis, the covering cuticle is thrown off along with the rest of the exoskeleton; while the rudimentary limb straightens out, and, though very small, acquires all the organization appropriate to that limb. At every moult it grows; but, it is only after a long time that it acquires nearly the size of its uninjured and older fellow. Hence, it not unfrequently happens, that crayfish are found with pincers and other limbs, which, though alike useful and anatomically complete, are very unequal in size.

Injuries inflicted while the crayfish are soft after moulting, are apt to produce abnormal growths of the part affected; and these may be perpetuated, and give rise to various monstrosities, in the pincers and in other parts of the body.

In the reproduction of their kind by means of eggs the co-operation of the males with the females is necessary. On the basal joint of the hindermost pair of legs of the male a small aperture is to be seen (fig. 3, A; *vd*). In these, the ducts of the apparatus in which the fecundating substance is formed terminate. The fecundating material itself is a thickish fluid, which sets into a white solid after extrusion. The male deposits this substance on the thorax of the female, between the bases of the hindermost pairs of thoracic limbs.

The eggs formed in the ovary are conducted to apertures, which are situated on the bases of the last pair of ambulatory legs but two, that is, in the hinder of the two pair which are provided with chelate extremities (fig. 3, B; *od*).

After the female has received the deposit of the spermatic matter of the male, she retires to a burrow, in the manner already stated, and then the process of laying the eggs commences. These, as they leave the apertures of the oviducts, are coated with a viscid matter, which is readily drawn out into a short thread. The end of the thread attaches itself to one of the long hairs, with which the swimmerets are fringed, and as the viscid matter rapidly hardens, the egg thus becomes attached to the limb by a stalk. The operation is repeated, until sometimes a couple of hundred eggs are thus glued on to the swimmerets. Partaking in the movements of the swimmerets, they are washed backwards and forwards in the water, and thus aërated and kept free of impurities; while the young crayfish is formed much in the same way as the chick is formed in a hen's egg.

The process of development, however, is very slow, as it occupies the whole winter. In late spring-time, or early summer, the young burst the thin shell of the egg, and, when they are hatched, present a general resemblance to their parents. This is very unlike what takes place in crabs and lobsters, in which the young leave the egg in a condition very different from the parent, and undergo a remarkable metamorphosis before they attain their proper form.

For some time after they are hatched, the young hold on to the swimmerets of the mother, and are carried about, protected by her abdomen, as in a kind of nursery.

That most careful naturalist, Roesel von Rosenhof, says of the young, when just hatched :—

"At this time they are quite transparent; and when

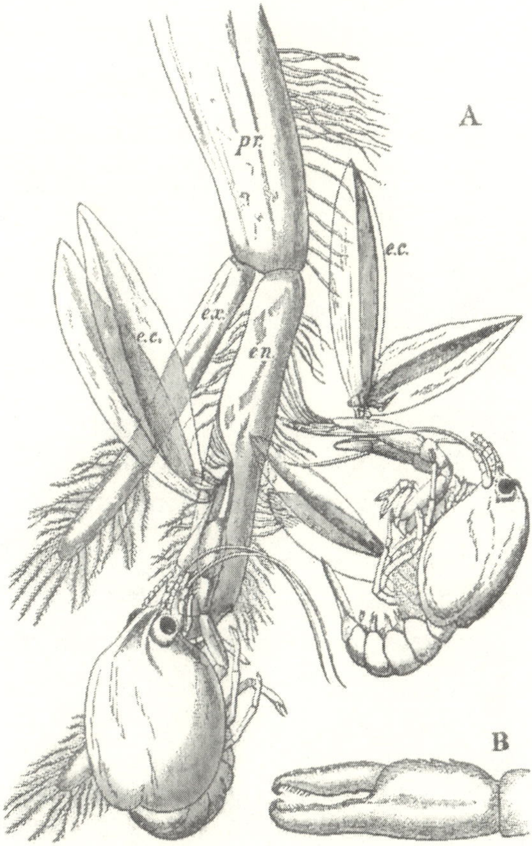

FIG. 8.—*Astacus fluviatilis.*—A, two recently hatched crayfish attached to one of the swimmerets of the mother (× 4). *pr*, protopodite; *en*, endopodite; and *ex*, exopodite of the swimmeret; *ec*, ruptured egg-cases. B, chela of a recently hatched crayfish (× 10).

such a crayfish [a female with young] is brought to table, it looks quite disgusting to those who do not know what the young are; but if we examine it more closely, especially with a magnifying-glass, we see with pleasure that the little crayfish are already perfect, and resemble the large one in all respects. When the mother of these little crayfish, after they have begun to be active, is quiet for a while, they leave her and creep about, a short way off. But, if they spy the least sign of danger, or there is any unusual movement in the water, it seems as if the mother recalled them by a signal; for they all at once swiftly return under her tail, and gather into a cluster, and the mother hies to a place of safety with them, as quickly as she can. A few days later, however, they gradually forsake her." *

Fishermen declare that "Hen Lobsters" protect their young in a similar manner.† Jonston,‡ who wrote in the middle of the seventeenth century, says that the little crayfish are often to be seen adhering to the tail of the mother. Roesel's observations imply the same thing; but he does not describe the exact mode of adherence, and I can find no observations on the subject in the works of later writers.

It has been seen that the eggs are attached to the swimmerets by a viscid substance, which is, as it were, smeared over them and the hairs with which they are

* "Der Monatlich-herausgegeben Insecten Belustigung." Dritter Theil, p. 336. 1755.

† Bell's "British Crustacea," p. 249.

‡ "Joannis Jonstoni Historiæ naturalis de Piscibus et Cetis Libri quinque. Tomus IV. 'De Cammaro seu Astaco fluviatili.'"

fringed, and is continued by longer or shorter thread-like pedicles into the coat of the same material which invests each egg. It very soon hardens, and then becomes very firm and elastic.

When the young crayfish is ready to be hatched, the egg case splits into two moieties, which remain attached, like a pair of watch glasses, to the free end of the pedicle of the egg (fig. 8, A; *ec*). The young animal, though very similar to the parent, does not quite "resemble it in all respects," as Roesel says. For not only are the first and the last pairs of abdominal limbs wanting, while the telson is very different from that of the adult; but the ends of the great chelæ are sharply pointed and bent down into abruptly incurved hooks, which overlap when the chelæ are shut (fig. 8, B). Hence, when the chelæ have closed upon anything soft enough to allow of the imbedding of these hooks, it is very difficult, if not impossible, to open them again.

Immediately the young are set free, they must instinctively bury the ends of their forceps in the hardened egg-glue which is smeared over the swimmerets, for they are all found to be holding on in this manner. They exhibit very little movement, and they bear rough shaking or handling without becoming detached; in consequence, I suppose, of the interlocking of the hooked ends of the chelæ imbedded in the egg-glue.

Even after the female has been plunged into alcohol, the young remain attached. I have had a female, with young affixed in this manner, under observation for five days, but none of them showed any signs of detaching themselves; and I am inclined to think that they are set free only at the first moult. After this, it would appear that the adhesion to the parent is only temporary.

The walking legs are also hooked at their extremities, but they play a less important part in fixing the young to the parent, and seem to be always capable of loosing their hold.

I find the young of a Mexican crayfish (*Cambarus*) to be attached in the same manner as those of the English crayfish; but, according to Mr. Wood-Mason's recent observations, the young of the New Zealand crayfishes fix themselves to the swimmerets of the parent by the hooked ends of their hinder ambulatory limbs.

Crayfishes, in every respect similar to those found in our English rivers, that is to say, of the species *Astacus fluviatilis*, are met with in Ireland, and on the Continent, as far south as Italy and northern Greece; as far east as western Russia; and as far north as the shores of the Baltic. They are not known to occur in Scotland; in Spain, except about Barcelona, they are either rare, or have remained unnoticed.

There is, at present, no proof of the occurrence of *Astacus fluviatilis* in the fossil state.

Curious myths have gathered about crayfishes, as about other animals. At one time "crabs'-eyes" were collected in vast numbers, and sold for medicinal purposes as a remedy against the stone, among other diseases. Their real utility, inasmuch as they consist almost entirely of carbonate of lime, with a little phosphate of lime and animal matter, is much the same as that of chalk, or carbonate of magnesia. It was, formerly, a current belief that crayfishes grow poor at the time of new moon, and fat at that of full moon; and, perhaps, there may be some foundation for the notion, considering the nocturnal habits of the animals. Van Helmont, a great dealer in wonders, is responsible for the story that, in Brandenburg, where there is a great abundance of crayfishes, the dealers were obliged to transport them to market by night, lest a pig should run under the cart. For if such a misfortune should happen, every crayfish would be found dead in the morning: "Tam exitialis est porcus cancro." Another author improves the story, by declaring that the steam of a pig-stye, or of a herd of swine, is instantaneously fatal to crayfish. On the other hand, the smell of putrifying crayfish, which is undoubtedly of the strongest, was said to drive even moles out of their burrows.

CHAPTER II.

THE PHYSIOLOGY OF THE CRAYFISH. THE MECHANISM BY WHICH THE PARTS OF THE LIVING ENGINE ARE SUPPLIED WITH THE MATERIALS NECESSARY FOR THEIR MAINTENANCE AND GROWTH.

AN analysis of such a sketch of the "Natural History of the Crayfish" as is given in the preceding chapter, shows that it provides brief and general answers to three questions. First, what is the form and structure of the animal, not only when adult, but at different stages of its growth? Secondly, what are the various actions of which it is capable? Thirdly, where is it found? If we carry our investigations further, in such a manner as to give the fullest attainable answers to these questions, the knowledge thus acquired, in the case of the first question, is termed the *Morphology* of the crayfish; in the case of the second question, it constitutes the *Physiology* of the animal; while the answer to the third question would represent what we know of its *Distribution* or *Chorology*. There remains a fourth problem, which can hardly be regarded as seriously under discussion, so long as knowledge has advanced no further than the Natural History stage; the question, namely, how all these facts comprised under Morphology, Physiology, and Chorology have come to be what they are; and the attempt to solve this problem leads us to the crown of Biological effort, *Ætiology*. When it supplies answers to all the questions which fall under these four heads, the Zoology of Crayfish will have said its last word.

As it matters little in what order we take the first three questions, in expanding Natural History into Zoology, we may as well follow that which accords with the history of science. After men acquired a rough and general knowledge of the animals about them, the next thing which engaged their interest was the discovery in these animals of arrangements by which results, of a kind similar to those which their own ingenuity effects through mechanical contrivances, are brought about. They observed that animals perform various actions; and, when they looked into the disposition and the powers of the parts by which these actions are performed, they found that these parts presented the characters of an apparatus, or piece of mechanism, the action of which could be deduced from the properties and connections of its constituents, just as the striking of a clock can be deduced from the properties and connections of its weights and wheels.

Under one aspect, the result of the search after the *rationale* of animal structure thus set afoot is *Teleology*; or the doctrine of adaptation to purpose. Under another aspect, it is *Physiology;* so far as Physiology consists in the elucidation of complex vital phenomena by deduction from the established truths of Physics and Chemistry, or from the elementary properties of living matter.

We have seen that the crayfish is a voracious and indiscriminate feeder; and we shall be safe in assuming that, if duly supplied with nourishment, a full-grown crayfish will consume several times its own weight of food in the course of the year. Nevertheless, the increase of the animal's weight at the end of that time is, at most, a small fraction of its total weight; whence it is quite clear, that a very large proportion of the food taken into the body must, in some shape or other, leave it again. In the course of the same period, the crayfish absorbs a very considerable quantity of oxygen, supplied by the atmosphere to the water which it inhabits; while it gives out, into that water, a large amount of carbonic acid, and a larger or smaller quantity of nitrogenous and other excrementitious matters. From this point of view, the crayfish may be regarded as a kind of chemical manufactory, supplied with certain alimentary raw materials, which it works up, transforms, and gives out in other shapes. And the first physiological problem which offers itself to us is the mode of operation of the apparatus contained in this factory, and the extent to which the products of its activity are to be accounted for by reasoning from known physical and chemical principles.

We have learned that the food of the crayfish is made up of very diverse substances, both animal and vegetable; but, so far as they are competent to nourish the animal permanently, these matters all agree in containing a peculiar nitrogenous body, termed *protein*, under one of its many forms, such as albumen, fibrin, and the like. With this may be associated fatty matters, starchy and saccharine bodies, and various earthy salts. And these, which are the essential constituents of the food, may be, and usually are, largely mixed up with other substances, such as wood, in the case of vegetable food, or skeletal and fibrous parts, in the case of animal prey, which are of little or no utility to the crayfish.

The first step in the process of feeding, therefore, is to reduce the food to such a state, that the separation of its nutritive parts, or those which can be turned to account, from its innutritious, or useless, constituents, may be facilitated. And this preliminary operation is the subdivision of the food into morsels of a convenient size for introduction into that part of the machinery in which the extraction of the useful products is performed.

The food may be seized by the pincers, or by the anterior chelate ambulatory limbs; and, in the former case, it is usually, if not always, transferred to the first, or second, or both of the anterior pairs of ambulatory limbs. These grasp the food, and, tearing it into pieces of the proper dimensions, thrust them between the external maxillipedes, which are, at the same time, worked rapidly to and fro sideways, so as to bring their toothed edges to bear upon the morsel. The other five pairs of jaws are no less active, and they thus crush and divide the food brought to them, as it is passed between their toothed edges to the opening of the mouth.

As the alimentary canal stretches from the mouth, at one end, to the vent at the other, and, at each of these limits, is continuous with the wall of the body, we may conceive the whole crayfish to be a hollow cylinder, the cavity of which is everywhere closed, though it is traversed by a tube, open at each end (fig. 6). The shut cavity between the tube and the walls of the cylinder may be termed the *perivisceral cavity;* and it is so much filled up by the various organs, which are interposed between the alimentary canal and the body wall, that all that is left of it is represented by a system of irregular channels, which are filled with blood, and are termed *blood sinuses.* The wall of the cylinder is the outer wall of the body itself, to which the general name of *integument* may be given; and the outermost layer of this, again, is the *cuticle,* which gives rise to the whole of the exoskeleton. This cuticle, as we have seen, is extensively impregnated with lime salts; and, moreover, in consequence of its containing *chitin,* it is often spoken of as the *chitinous cuticula.*

Having arrived at this general conception of the disposition of the parts of the factory, we may next proceed to consider the machinery of alimentation which is contained within it, and which is represented by the various divisions of the alimentary canal, with its appendages; by the apparatus for the distribution of nutriment; and by two apparatuses for getting rid of those products which are the ultimate result of the working of the whole organism.

And here we must trench somewhat upon the province of *Morphology*, as some of these pieces of apparatus are complicated; and their action cannot be comprehended without a certain knowledge of their anatomy.

The mouth of the crayfish is a longitudinally elongated, parallel-sided opening, in the integument of the ventral or sternal aspect of the head. Just outside its lateral boundaries, the strong mandibles project, one on each side (fig 3, B; *4*); their broad crushing surfaces, which are turned towards one another, are therefore completely external to the oral cavity. In front, the mouth is overlapped by a wide shield-shaped plate termed the upper lip, or *labrum* (figs. 3 and 6, *lb*); while, immediately behind the mandibles, there is, on each side, an elongated fleshy lobe, joined with its fellow by the posterior boundary of the mouth. These together constitute the *metastoma* (fig. 3, B; *mt*), which is sometimes called the lower lip. A short wide gullet, termed the œsophagus (fig. 6, *oe*), leads directly upwards into a spacious bag, the *stomach*, which occupies almost the whole cavity of the head. It is divided by a constriction into a large anterior chamber (*cs*), into the under face of which the gullet opens, and a small posterior chamber (*ps*), from which the intestine (*hg*) proceeds.

In a man's stomach, the opening by which the gullet communicates with the stomach is called the *cardia*, while that which places the stomach in communication with the intestine is named the *pylorus*; and these terms having been transferred from human anatomy to that of the lower animals, the larger moiety of the crayfish's stomach is called the *cardiac division*, while the smaller is termed the *pyloric division* of the organ. It must be recollected, however, that, in the crayfish, the so-called cardiac division is that which is actually furthest from the heart, not that which is nearest to it, as in man.

The gullet is lined by a firm coat which resembles thin parchment. At the margins of the mouth, this strong lining is easily seen to be continuous with the cuticular exoskeleton; while, at the cardiac orifice, it spreads out and forms the inner or cuticular wall of the whole gastric cavity, as far as the pylorus, where it ends in certain valvular projections. The chitinous cuticle which forms the outermost layer of the integument is thus, as it were, turned in, to constitute the innermost layer of the walls of the stomach; and it confers upon them so great an amount of stiffness that they do not collapse when the organ is removed from the body. Furthermore, just as the cuticle of the integument is calcified to form the hard parts of the exoskeleton, so is the cuticle of the stomach calcified, or otherwise hardened, to give rise, in the first place, to the very remarkable and complicated apparatus which has already been spoken of, as a sort of *gastric mill*

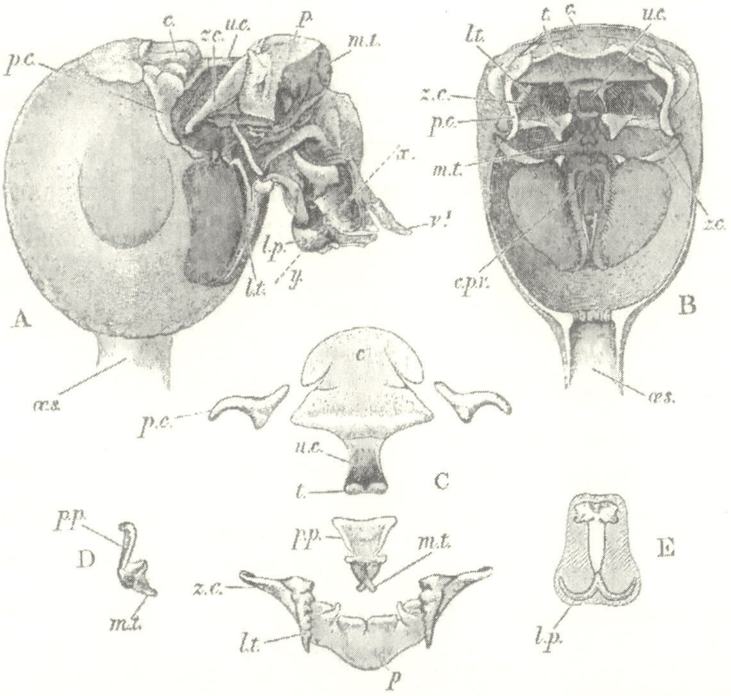

FIG. 9.—*Astacus fluviatilis.*—A, the stomach with its outer coat removed, seen from the left side; B, the same viewed from the front, after removal of the anterior wall; C, the ossicles of the gastric mill separated from one another; D, the prepyloric ossicle and median tooth, seen from the right side; E, transverse section of the pyloric region along the line *xy* in A (all × 2). *c*, cardiac ossicle; *cpv*, cardio-pyloric valve; *lp*, lateral pouch; *lt*, lateral tooth, seen through the wall of the stomach in A; *mg*, mid-gut; *mt*, median tooth, seen through the wall of the stomach in A; *œs*, œsophagus; *p*, pyloric ossicle; *pc*, pterocardiac ossicle; *pp*, prepyloric ossicle; *uc*, uro-cardiac process; *t*, convexities on the free surface of its hinder end; v^1, median pyloric valve; *zc*, zygocardiac ossicle.

or *food-crusher;* and, secondly, to a *filter* or *strainer*, whereby the nutritive juices are separated from the innutritious hard parts of the food and passed on into the intestine.

The gastric mill begins in the hinder half of the cardiac division. Here, on the upper wall of the stomach, we see a broad transverse calcified bar (figs. 9–11, *c*) from the middle of the hinder part of which another bar (*uc*), united to the first by a flexible portion, is continued backwards in the middle line. The whole has, therefore, somewhat the shape of a cross-bow. Behind the first-mentioned piece, the dorsal wall of the stomach is folded in, in such a manner as to give rise to a kind of pouch; and the second piece, or what we may call the handle of the crossbow, lies in the front wall of this pouch. The end of this piece is dense and hard, and its free surface, which looks into the top of the cardiac chamber, is raised into two oval, flattened convex surfaces (*t*). Connected by a transverse joint with the end of the handle of the crossbow, there is another solid bar, which ascends obliquely forwards in the back wall of the pouch (*pp*). The end which is articulated with the handle of the crossbow is produced into a strong reddish conical tooth (*mt*), curved forwards and bifurcated at the summit; consequently, when the cavity of the stomach is inspected from the fore part of the cardiac pouch (fig. 9, B), the two-pointed curved tooth (*mt*) is seen projecting behind the convex surfaces (*t*), in the middle line, into the interior of that cavity. The joint which connects the handle of the crossbow with the hinder middle piece is elastic; hence, if the two are straightened out, they return to their bent disposition as soon as they are released. The upper end of the hinder middle piece (*pp*) is connected with a second flat transverse plate which lies in the dorsal wall of the pyloric chamber (*p*). The whole arrangement, thus far, may be therefore compared to a large cross-bow and a small one, with the ends of their handles fastened together by a spring joint, in such a manner that the handle of the one makes an acute angle with the handle of the other; while the middle of each bow is united with the middle of the other by the bent arm formed by the two handles. But, in addition to this, the outer ends of the two bows are also connected together. A small, curved, calcified bar (*pc*) passes from the outer end of the front crosspiece downwards and outwards in the wall of the stomach, and its hinder and lower extremity is articulated with another larger bar (*zc*) which runs upwards and backwards to the hinder or pyloric crosspiece, with which it articulates. Internally, this piece projects into the cardiac cavity of the stomach as a stout elongated reddish elevation (*lt*), the surface of which is produced into a row of strong sharp, transverse ridges, which diminish in size from before backwards, and constitute a crushing surface almost like that of the grinder of an elephant. Thus, when the front part of the cardiac cavity is cut away, not only are the median teeth already mentioned seen, but, on each side of them, there is one of these long lateral teeth.

There are two small pointed teeth, one under each of the lateral teeth, and each of these is supported by a broad plate, hairy on its inner surface, which enters into the lateral wall of the cardiac chamber. There are various other smaller skeletal parts, but the most im-

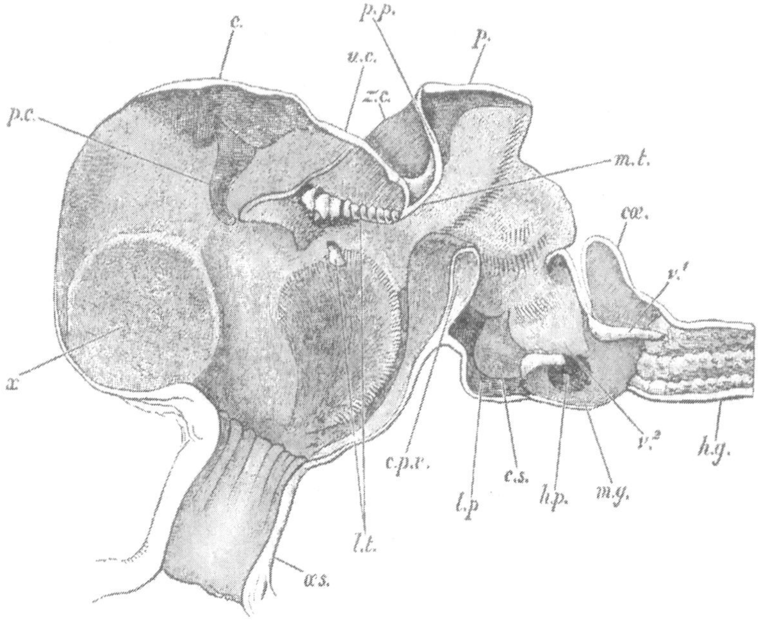

FIG. 10.—*Astacus fluviatilis.*—Longitudinal section of the stomach (× 4). *c*, cardiac ossicle; *cæ*, cæcum; *c.p.v*, cardio-pyloric valve; *cs*, cushion-shaped surface; *hg*, hind-gut; *hp*, aperture of right bile duct; *lp*, lateral pouch; *lt*, lateral teeth; *mg*, mid-gut; *mt*, median tooth; *œs*, œsophagus; *p*, pyloric ossicle; *pc*, pterocardiac ossicles; *pp*, prepyloric ossicle; *uc*, urocardiac process; v^1, median pyloric valve; v^2, lateral pyloric valve; *x*, position of gastrolith; *zc*, zygocardiac ossicle.

portant are those which have been described; and these, from what has been said, will be seen to form a sort of hexagonal frame, with more or less flexible joints at the angles, and having the anterior and the posterior sides connected by a bent jointed middle bar. As all these parts are merely modifications of the hard skeleton, the apparatus is devoid of any power of moving itself. It is set in motion, however, by the same substance as that which gives rise to all the other bodily movements of the crayfish, namely, *muscle*. The chief muscles which move it are four very strong bundles of fibres. Two of these are attached to the front crosspiece, and proceed thence, upwards and forwards, to be fixed to the inner face of the carapace in the front part of the head (figs. 5, 6, and 12, *ag*). The two others, which are fixed into the hinder crosspiece and hinder lateral pieces, pass upwards and backwards, to be attached to the inner face of the carapace in the back part of the head (*pg*). When these muscles shorten, or contract, they pull the front and back crosspieces further away from one another; consequently, the angle between the handles becomes more open and the tooth which is borne on their ends travels downwards and forwards. But, at the same time, the angle between the side bars becomes more open and the lateral tooth of each side moves inwards till it crosses in front of the middle tooth, and strikes against this and the opposite lateral tooth, which has undergone a corresponding change of place. The muscles being now relaxed, the elasticity of the joints suffices to bring the whole apparatus back to its first position, when a new contraction brings about a new clashing of the teeth. Thus, by the alternate contraction and relaxation of these two pair of muscles, the three teeth are made to stir up and crush whatever is contained in the cardiac chamber. When the stomach is removed and the front part of the cardiac chamber is cut away, the front cross-piece may be seized with one pair of forceps and the hind cross-piece with another. On slightly pulling the two, so as to imitate the action of the muscles, the three teeth will be found to come together sharply, exactly in the manner described.

Works on mechanics are full of contrivances for the conversion of motion; but it would, perhaps, be difficult to discover among these a prettier solution of the problem; given a straight pull, how to convert it into three simultaneous convergent movements of as many points.

What I have called the *filter* is constructed mainly out of the chitinous lining of the pyloric chamber. The aperture of communication between this and the cardiac chamber, already narrow, on account of the constriction of the walls of the stomach at this point, is bounded at the sides by two folds; while, from below, a conical tongue-shaped process (figs. 6, 10, and 11, *cpv*), the surface of which is covered with hairs, further obstructs the opening. In the posterior half of the pyloric chamber, its side walls are, as it were, pushed in; and, above, they so nearly meet in the middle line, that a mere vertical chink is left between them; while even this is crossed by hairs set upon the two surfaces. In its lower half, however, each side wall curves outwards, and forms a cushion-shaped surface (fig. 10, *cs*) which looks downwards and inwards. If the floor of the pyloric chamber were flat, a wide triangular passage would thus be left open in its lower half. But, in fact, the floor rises into a ridge in the middle, while, at the sides, it adapts itself to the shape of the two cushion-shaped surfaces; the result of which is that the whole cavity of the posterior part of the pyloric division of the stomach is reduced to a narrow three-rayed fissure. In transverse section, the vertical ray of this fissure is straight, while the two lateral ones are concave upwards (fig. 9, *E*). The cushions of the side walls are covered with short close-set hairs. The corresponding surfaces of the floor are raised into longitudinal parallel ridges, the edge of each of which is fringed with very fine hairs. As everything which passes from the cardiac sac to the intestine must traverse this singular apparatus, only the most finely divided solid matters can escape stoppage, so long as its walls are kept together.

Finally, at the opening of the pyloric sac into the intestine, the chitinous investment terminates in five symmetrically arranged processes, the disposition of which is such that they must play the part of valves in preventing any sudden return of the contents of the intestine to the stomach, while they readily allow of a passage the other way. One of these valvular processes is placed in the middle line above (figs. 10 and 11, v^1). It is longer than the others and concave below. The lateral processes (v^2,) of which there are two on each side, are triangular and flat.

The cuticular lining which gives rise to all the complicated apparatus which has just been described, must

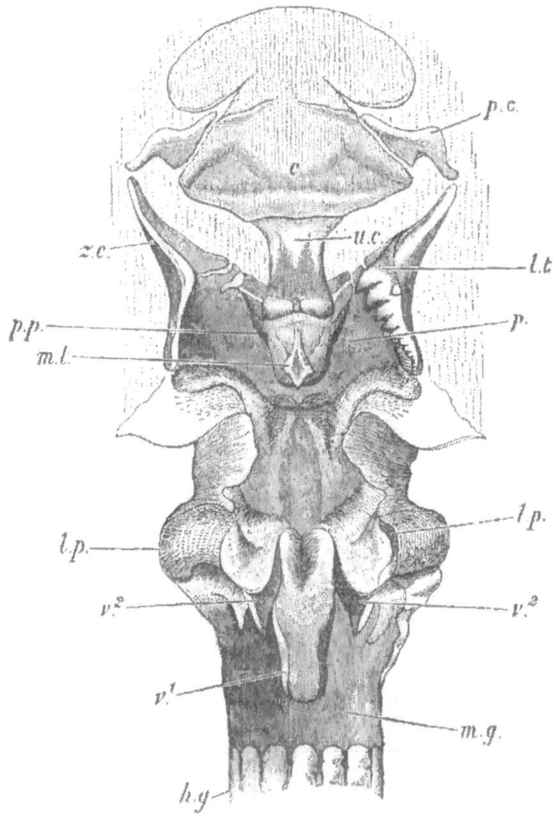

FIG. 11.—*Astacus fluviatilis.*—View of the roof of the stomach, the ventral wall of which, and of the mid-gut, is laid open by a longitudinal incision (× 4). On the right side (the left in the figure), the lateral tooth is cut away, as well as the floor of the lateral pouch. The letters have the same signification as in fig. 10.

not be confounded with the proper wall of the stomach, which invests it, and to which it owes it origin, just as the cuticle of the integument is produced by the soft true skin which lies beneath it. The wall of the stomach is a soft pale membrane containing variously disposed muscular fibres; and, beyond the pylorus, it is continued into the wall of the intestine.

It has already been mentioned that the intestine is a slender and thin-walled tube, which passes straight through the body almost without change, except that it becomes a little wider and thicker-walled near the vent. Immediately behind the pyloric valves, its surface is quite smooth and soft (figs. 9, 10, and 12, *mg*), and its floor presents a relatively large aperture, the termination of the bile duct (fig. 12, *bd*, fig. 10, *hp*.), on each side. The roof is, as it were, pushed out into a short median pouch or *cæcum* (*cæ*). Behind this, its character suddenly changes, and six squarish elevations, covered with a chitinous cuticle, encircle the cavity of the intestine (*r*). From each of these, a longitudinal ridge, corresponding with a fold of the wall of the intestine, takes its rise, and passes, with a slight spiral twist, to its extremity (*hg*). Each of these ridges is beset with small papillæ, and the chitinous lining is continued over the whole to the vent, where it passes into the general cuticle of the integument, just as the lining of the stomach is continuous with the cuticle of the integument at the mouth. The alimentary canal may, therefore, be distinguished into a *fore* and a *hind-gut* (*hg*), which have a thick internal lining of cuticular membrane; and a very short *mid-gut* (*mg*), which has no thick cuticular layer. It will be of

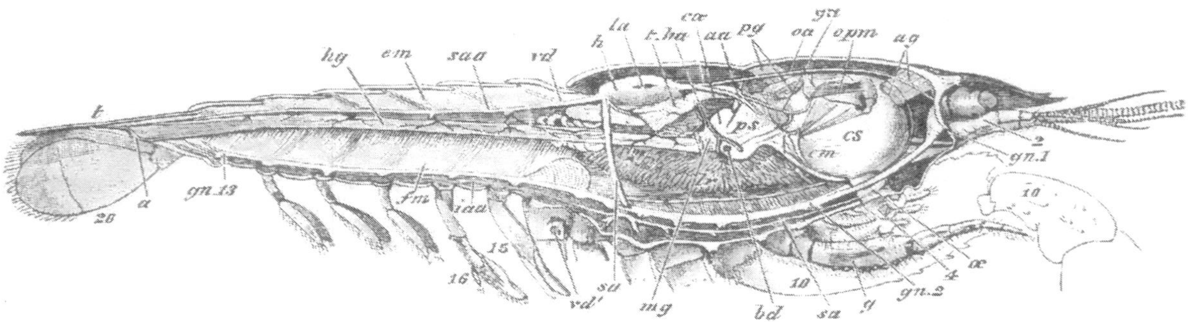

FIG. 12.—*Astacus fluviatilis.*—A dissection of a male specimen from the right side (nat. size). *a*, anus ; *aa*, antennary artery, cut short ; *ag*, anterior gastric muscles, the right cut away to its insertion ; *bd*, aperture of right bile duct ; *cm*, constrictor muscles of stomach ; *cæ*, cæcum ; *cpm*, right cardio-pyloric muscle ; *cs*, cardiac portion of stomach ; *em*, extensor muscles of abdomen ; *fm*, flexor muscles of abdomen ; *ga*, gastric artery ; *gn. 1*, supracesophageal ganglion ; *gn. 2*, sub-œsophageal ganglion ; *gn. 13*, last abdominal ganglion ; *h*, heart ; *ha*, hepatic artery ; *hg*, hind-gut ; *iaa*, inferior abdominal artery ; *la*, right lateral aperture of heart ; *lr*, left liver ; *mg*, mid-gut ; *oa*, ophthalmic artery ; *œ*, œsophagus ; *pg*, posterior gastric muscles, the right cut away to its insertion ; *ps*, pyloric portion of stomach ; *sa*, sternal artery ; *saa*, superior abdominal artery ; *t* (to the left), telson ; *t* (near the heart), testis ; *vd*, left vas deferens ; *vd'*, aperture of left vas deferens ; *2*, right antennule ; *4*, left mandible ; *9*, left external maxillipede ; *10*. left forceps ; *15*, first, *16*, second, and *20*, sixth abdominal appendages of the left side.

importance to recollect this distinction by-and-by, when the development of the alimentary canal is considered.

If the treatment to which the food is subjected in the alimentary apparatus were of a purely mechanical nature, there would be nothing more to describe in this part of the crayfish's mechanism. But, in order that the nutritive matters may be turned to account, and undergo the chemical metamorphoses, which eventually change them into substances of a totally different character, they must pass out of the alimentary canal into the blood. And they can do this only by making their way through the walls of the alimentary canal; to which end they must either be in a state of extremely fine division, or they must be reduced to the fluid condition. In the case of the fatty matters, minute subdivision may suffice; but the amylaceous substances and the insoluble protein compounds, such as the fibrin of flesh, must be brought into a state of solution. Therefore some substances must be poured into the alimentary canal, which, when mixed with the crushed food, will play the part of a chemical agent, dissolving out the insoluble proteids, changing the amyloids into soluble sugar, and converting all the proteids into those diffusible forms of protein matter, which are known as *peptones*.

The details of the processes here indicated, which may be included under the general name of *digestion*, have only quite recently been carefully investigated in the crayfish; and we have probably still much to learn about them; but what has been made out is very interesting, and proves that considerable differences exist between crayfishes and the higher animals in this respect.

The physiologist calls those organs, the function of which is to prepare and discharge substances of a special character, *glands;* and the matter which they elaborate is termed their *secretion.* On the one side, glands are in relation with the blood, whence they derive the materials which they convert into the substances characteristic of their secretion; on the other side, they have access, directly or indirectly, to a free surface, on to which they pour their secretion as it is formed.

Of such glands, the alimentary canal of the crayfish is provided with a pair, which are not only of very large size, but are further extremely conspicuous, on account of their yellow or brown colour. These two glands (figs. 12 and 13, *lr*) are situated beneath, and on each side of, the stomach and the anterior part of the intestine, and answer in position to the glands termed liver and pancreas in the higher animals, inasmuch as they pour their secretion into the mid-gut. These glands have hitherto always been regarded as the *liver,* and the name may be retained, though their secretion appears rather to correspond with the pancreatic fluid than with the bile of the higher animals.

Each liver consists of an immense number of short tubes, or *cæca,* which are closed at one end, but open at the other into a general conduit, which is termed their *duct.* The mass of the liver is roughly divided into

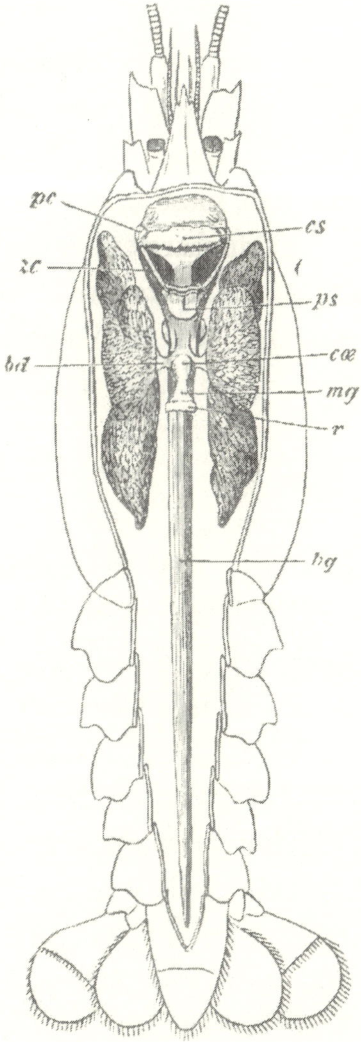

FIG. 13.—*Astacus fluviatilis.*—The alimentary canal and livers seen from above (nat. size). *bd*, bile-duct ; *cæ*, cæcum ; *cs*, cardiac portion of stomach, the line pointing to the cardiac ossicle ; *hg*, hind-gut ; *mg*, mid-gut ; *pc*, pterocardiac ossicle ; *ps*, pyloric portion of stomach, the line pointing to the pyloric ossicle ; *r*, ridge separating mid-gut from hind-gut ; *zc*, zygocardiac ossicle.

three lobes, one anterior, one lateral, and one posterior; and each lobe has its main duct, into which all the tubes composing it open. The three ducts unite together into a wide common duct (*bd*), which opens, just behind the pyloric valves, into the floor of the mid-gut. Hence the apertures of the two *hepatic ducts* are seen, one on each side, in this part of the alimentary canal when it is laid open from above. Every cæcum of the liver has a thin outer wall, lined internally by a layer of cells, constituting what is termed an *epithelium;* and, at the openings of the hepatic ducts, this epithelium passes into a layer of somewhat similar structure, which lines the mid-gut, and is continued through the rest of the alimentary canal, beneath the cuticula. Hence the liver may be regarded as a much divided side pouch of the mid-gut.

The epithelium is made up of *nucleated cells*, which are particles of simple living matter, or *protoplasm*, in the midst of each of which is a rounded body, which is termed the *nucleus*. It is these cells which are the seat of the manufacturing process which results in the formation of the secretion; it is, as it were, their special business to form that secretion. To this end they are constantly being newly formed at the summits of the cæca. As they grow, they pass down towards the duct and, at the same time, separate into their interior certain special products, among which globules of yellow fatty matter are very conspicuous. When these products are fully formed, what remains of the substance of the cells dissolves away, and the yellow fluid accumulating in the ducts passes into the mid-gut. The yellow colour is due to the globules of fat. In the young cells, at the summit of the cæca, these are either absent, or very small, whence the part appears colourless. But, lower down, small yellow granules appear in the cells, and these become bigger and more numerous in the middle and lower parts. In fact, few glands are better fitted for the study of the manner in which secretion is effected than the crayfish's liver.

We may now consider the alimentary machinery, the general structure of which has been explained, in action.

The food, already torn and crushed by the jaws, is passed through the gullet into the cardiac sac, and there reduced to a still more pulpy state by the gastric mill. By degrees, such parts as are sufficiently fluid are drained off into the intestine, through the pyloric strainer, while the coarser parts of the useless matters are probably rejected by the mouth, as a hawk or an owl rejects his casts. There is reason to believe, though it is not certainly known, that fluids from the intestine mix with the food while it is undergoing trituration, and effect the transformation of the starchy and the insoluble protein compounds into a soluble state. At any rate, as soon as the strained-off fluid passes into the mid-gut it must be mixed with the secretion of the liver, the action of which is probably similar to that of the pancreatic juice of the higher animals.

The mixture thus produced, which answers to the chyle of the higher animals, passes along the intestine, and the greater part of it, transuding through the walls of the alimentary canal, enters the blood, while the rest accumulates as dark coloured fæces in the hind gut, and

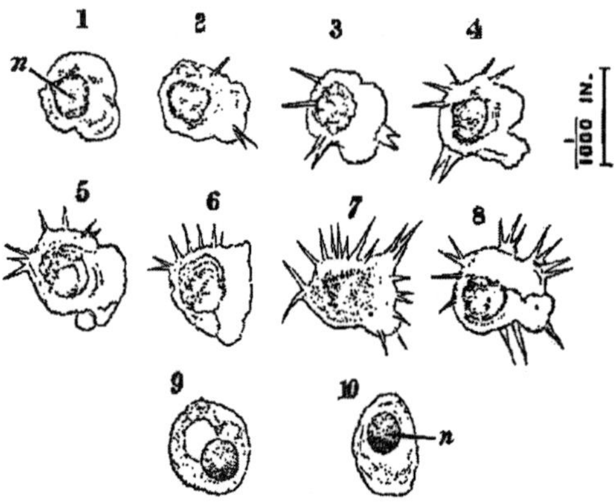

FIG. 14.—*Astacus fluviatilis.*—The corpuscles of the blood (highly magnified). *1-8* show the changes undergone by a single corpuscle during a quarter of an hour; *9* and *10* are corpuscles killed by magenta, and having the nucleus deeply stained by the colouring matter. *n*, nucleus.

is eventually passed out of the body by the vent. The fæcal matters are small in amount, and the strainer is so efficient that they rarely contain solid particles of sensible size. Sometimes, however, there are a good many minute fragments of vegetable tissue.

The blood of which the nutritive elements of the food

have thus become integral parts, is a clear fluid, either colourless, or of a pale neutral tint or reddish hue, which, to the naked eye, appears like so much water. But if subjected to microscopic examination, it is found to contain innumerable pale, solid particles, or *corpuscles*, which, when examined fresh, undergo constant changes of form (fig. 14). In fact, they correspond very closely with the colourless corpuscles which exist in our own blood; and, in its general characters, the crayfish's blood is such as ours would be if it were somewhat diluted and deprived of its red corpuscles. In other words, it resembles our lymph more than it does our blood. Left to itself it soon coagulates, giving rise to a pretty firm clot.

The sinuses, or cavities in which the greater part of the blood is contained, are disposed very irregularly in the intervals between the internal organs. But there is one of especially large size on the ventral or sternal side of the thorax (fig. 15, *sc*), into which all the blood in the body sooner or later makes its way. From this *sternal sinus* passages (*av*) lead to the gills, and from these again six canals (*bcv*), pass up on the inner side of the inner wall of each branchial chamber to a cavity situated in the dorsal region of the thorax, termed the *pericardium* (*p*), into which they open.

The blood of the crayfish is kept in a state of constant circulating motion by a pumping and distributing machinery, composed of the *heart* and of the *arteries*, with

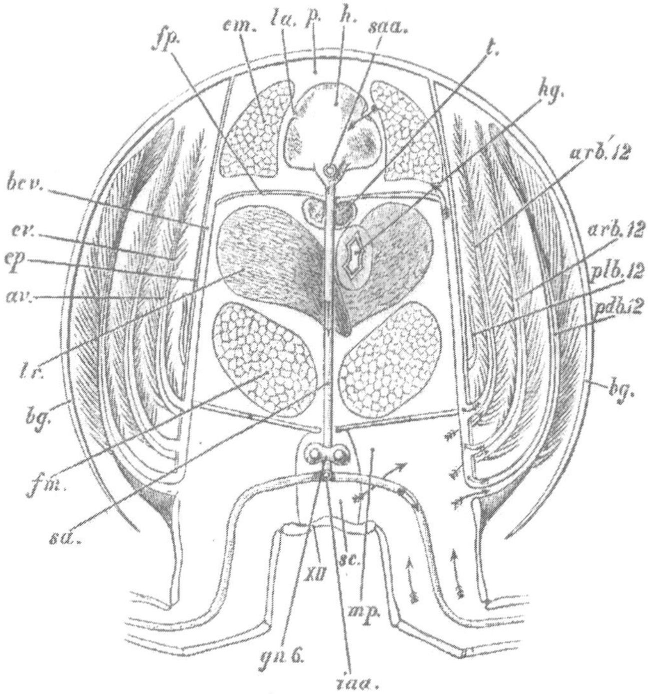

FIG. 15.—*Astacus fluviatilis.*—A diagrammatic transverse section of the thorax through the twelfth somite, showing the course of the circulation of the blood (× 3). *arb. 12*, the anterior or lower, and *arb'.* 12, the posterior or upper arthrobranchia of the twelfth somite; *av*, afferent branchial vessel; *bcv*, branchio-cardiac vein; *bg*, branchiostegite; *em*, extensor muscles of abdomen; *ep*, epimeral wall of thoracic cavity; *ev*, efferent branchial vessel; *fm*, flexor muscles of abdomen; *fp*, floor of pericardium; *gn. 6*, fifth thoracic ganglion; *h*, heart; *hg*, hind-gut; *iaa*, inferior abdominal artery, in cross section; *la*, lateral valvular apertures of heart; *lr*, liver; *mp*, indicates the position of the mesophragm by which the sternal canal is bounded laterally; *p*, pericardial sinus; *pdb. 12*, podobranchia, and *plb. 12*, pleurobranchia of the twelfth somite; *sa*, sternal artery; *saa*, superior abdominal artery; *sc*, sternal canal; *t*, testis; XII., sternum of twelfth somite. The arrows indicate the direction of the blood flow.

their larger and smaller branches, which proceed from it and ramify through the body, to terminate eventually in the blood sinuses, which represent the veins of the higher animals.

When the carapace is removed from the middle of the region which lies behind the cervical groove, that is, when the dorsal or *tergal* wall of the thorax is taken away, a spacious chamber is laid open which is full of blood. This is the cavity already mentioned as the *pericardium* (fig. 15, *p*), though, as it differs in some respects from that which is so named in the higher animals, it will be better to term it the *pericardial sinus*.

The heart (fig. 15, *h*), lies in the midst of this sinus. It is a thick muscular body (fig. 16), with an irregularly hexagonal contour when viewed from above, one angle of the hexagon being anterior and another posterior. The lateral angles of the hexagon are connected by bands of fibrous tissue (*ac*) with the walls of the pericardial sinus. Otherwise, the heart is free, except in so far as it is kept in place by the arteries which leave it and traverse the walls of the pericardium. One of these arteries (figs. 5, 12, and 16, *saa*), starting from the hinder part of the heart, of which it is a sort of continuation, runs along the middle line of the abdomen above the intestine, to which it gives off many branches. A second large artery starts from a dilatation, which is common to it with the foregoing, but passing directly downwards (figs. 12 and 15, *sa*, and fig. 16, *st. a*), either on the right or on the left side of the intestine, traverses the nervous cord (figs. 12 and 15), and divides into an anterior (fig. 12, *sa*) and a posterior (*iaa*) branch, both of which run beneath and parallel with that cord.

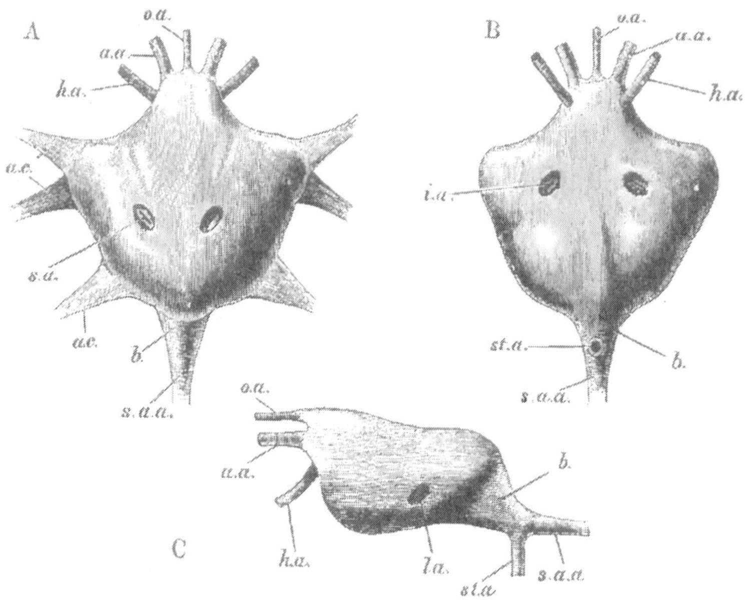

FIG 16.—*Astacus fluviatilis.*—The heart (× 4). A, from above; B, from below; C, from the left side. *aa*, antennary artery; *ac*, alæ cordis, or fibrous bands connecting the heart with the walls of the pericardial sinus; *b*, bulbous dilatation at the origin of the sternal artery; *ha*, hepatic artery; *la*, lateral valvular apertures; *oa*, ophthalmic artery; *s.a*, superior valvular apertures; *s.a.a*, superior abdominal artery; *st.a*, sternal artery, in B cut off close to its origin.

A third artery runs, from the front part of the heart, forwards in the middle line, over the stomach, to the eyes and fore part of the head (figs. 5, 12, and 16, *oa*); and two others diverge one on each side of this, and sweep round the stomach to the antennæ (*aa*). Behind these, yet two other arteries are given off from the under side of the heart, and supply the liver (*ha*). All these arteries branch out and eventually terminate in fine, so-called *capillary*, ramifications.

In the dorsal wall of the heart two small oval apertures are visible, provided with valvular lips (fig. 16, *sa*), which open inwards, or towards the internal cavity of the heart. There is a similar aperture in each of the lateral faces of the heart (*la*), and two others in its inferior face (*ia*), making six in all. These apertures readily admit fluid into the heart, but oppose its exit. On the other hand, at the origins of the arteries, there are small valvular folds, directed in such a manner as to permit the exit of fluid from the heart, while they prevent its entrance.

The walls of the heart are muscular, and, during life, they contract at intervals with a regular rhythm, in such a manner as to diminish the capacity of the internal cavity of the organ. The result is, that the blood which it contains is driven into the arteries, and necessarily forces into their smaller ramifications an equivalent amount of the blood which they already contained; whence, in the long run, the same amount of blood passes out of the ultimate capillaries into the blood sinuses. From the disposition of the blood sinuses, the impulse thus given to the blood which they contain is finally conveyed to the blood in the branchiæ, and a proportional quantity of that blood leaves the branchiæ and passes into the sinuses which connect them with the pericardial sinus (fig. 15, *bcv*), and thence into that cavity. At the end of the contraction, or *systole*, of the heart, its volume is of course diminished by the volume of the blood forced out, and the space between the walls of the heart and those of the pericardial sinus is increased to the same extent. This space, however, is at once occupied by the blood from the branchiæ, and perhaps by some blood which has not passed through the branchiæ, though this is doubtful. When the systole is over, the *diastole* follows; that it to say, the elasticity of the walls of the heart and that of the various parts which connect it with the walls of the pericardium, bring it back to its former size, and the blood in the pericardial sinus flows into its cavity by the six apertures. With a new systole the same process is repeated, and thus the blood is driven in a circular course through all parts of the body.

It will be observed that the branchiæ are placed in the course of the current of blood which is returning to the heart; which is the exact contrary of what happens in fishes, in which the blood is sent from the heart to the branchiæ, on its way to the body. It follows, from this arrangement, that the blood which goes to the branchiæ is blood in which the quantity of oxygen has undergone a diminution, and that of carbonic acid an increase, as compared with the blood in the heart itself. For the activity of all the organs, and especially of the muscles, is inseparably connected with the absorption of oxygen and the evolution of carbonic acid; and the only source from which the one can be derived, and the only receptacle into which the other can be poured, is the blood which bathes and permeates the whole fabric to which it is distributed by the arteries.

The blood, therefore, which reaches the branchiæ has lost oxygen and gained carbonic acid; and these organs constitute the apparatus for the elimination of the injurious gas from the economy on the one hand, and, on the other, for the taking in of a new supply of the needful "vital air," as the old chemists called it. It is thus that the branchiæ subserve the respiratory function.

The crayfish has eighteen perfect and two rudimentary branchiæ in each branchial chamber, the boundaries of which have been already described.

Of the eighteen perfect branchiæ, six (*podobranchiæ*) are attached to the basal joints of the thoracic limbs, from the last but one to the second (second maxillipede) inclusively (fig. 4, p. 26, *pdb*, and fig. 17, A, B); and eleven (*arthrobranchiæ*) are fixed to the flexible interarticular membranes, which connect these basal joints with the parts of the thorax to which they are articulated (fig. 4, *arb*, *arb'*, fig. 17, C). Of these eleven branchiæ, two are attached to the interarticular membranes of all the ambulatory legs but the last, (=6) and to those of the pincers and of the external maxillipedes, (=4) and one to that of the

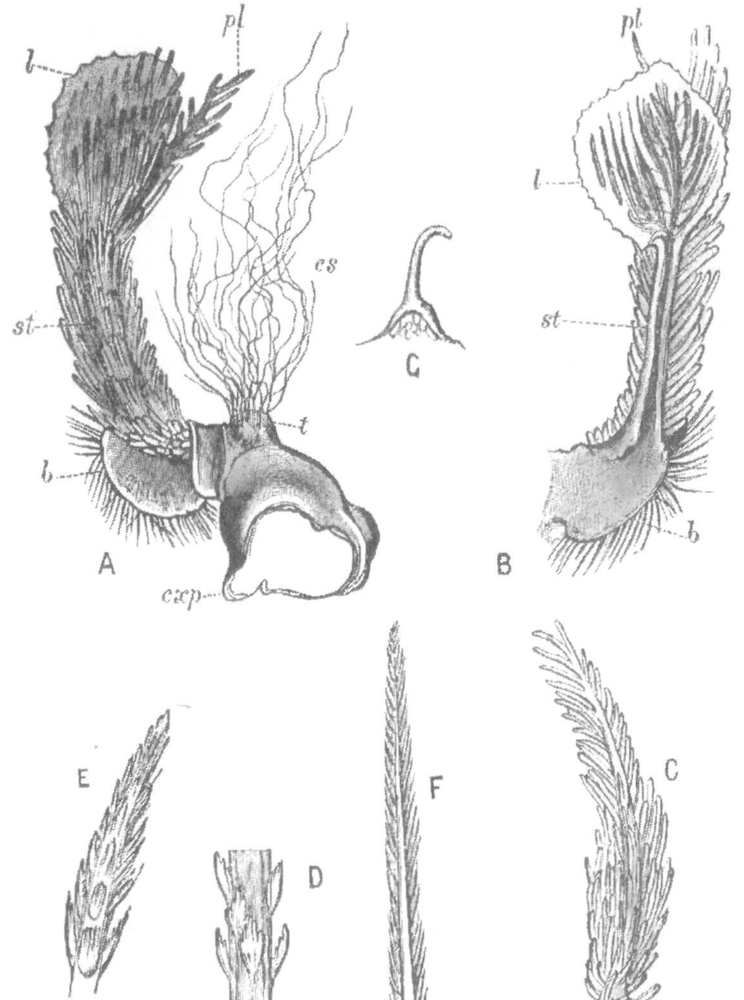

FIG. 17.—*Astacus fluviatilis.*—A, one of the podobranchiæ from the outer side; B, the same from the inner side; C, one of the arthrobranchiæ; D, a part of one of the coxopoditic setæ; E, extremity of the same seta; F, extremity of a seta from the base of the podobranchia; G, hooked seta of the lamina; (A—C, × 3; D—G, highly magnified). *b*, base of podobranchia; *cs*, coxopoditic setæ; *cxp*, coxopodite; *l*, lamina, *pl*, plume, and *st*, stem of podobranchia; *t*, tubercle on the coxopodite, to which the setæ are attached.

second maxillipede. The first maxillipede and the last ambulatory limb have none. Moreover, where there are two arthrobranchiæ, one is more or less in front of and external to the other.

These eleven arthrobranchiæ are all very similar in structure (fig. 17, C). Each consists of a stem which contains two canals, one external and one internal, separated by a longitudinal partition. The stem is beset with a great number of delicate *branchial filaments*, so that it looks like a plume tapering from its base to its summit. Each filament is traversed by large vascular channels, which break up into a net-work immediately beneath the surface. The blood driven into the external canals of the stem (fig. 15, *av*) is eventually poured into the inner canal (*ev*), which again communicates with the channels (*bcv*) which lead to the pericardial sinus (*p*). In its course, the blood traverses the branchial filaments, the outer investment of each of which is an excessively thin chitinous membrane, so that the blood contained in them is practically separated by a mere film from the aërated water in which the gills float. Hence, an exchange of gaseous constituents readily takes place, and as much oxygen is taken in as carbonic acid is given out.

The six podobranchiæ, or gills which are attached to the basal joints of the legs, play the same part, but differ a good deal in the details of their structure from those which are fixed to the interarticular membranes. Each consists of a broad *base* (fig. 17, A and B; *b*) beset with many fine straight hairs, or *setæ* (F), whence a narrow *stem* (*st*) proceeds. At its upper end this stem divides into two parts, that in front, the *plume* (*pl*), resembling the free end of one of the gills just described, while that behind, the *lamina* (*l*), is a broad thin plate, bent upon itself longitudinally in such a manner that its folded edge lies forwards, and covered with minute hooked setæ (G). The gill which follows is received into the space included between the two lobes or halves of the folded lamina (fig. 4, p. 26). Each lobe is longitudinally plaited into about a dozen folds. The whole front and outer face of the stem is beset with branchial filaments; hence, we may compare one of these branchiæ to one of the preceding kind, in which the stem has become modified and has given off a large folded lamina from its inner and posterior face.

The branchiæ now described are arranged in sets of three for each of the thoracic limbs, from the third maxillipede to the last but one ambulatory limb, and two for the second maxillipede, thus making seventeen in all ($3 \times 5 + 2 = 17$); and, between every two there is found a bundle of long twisted hairs (fig. 17, A, *cx.s*; D and E), which are attached to a small elevation (*t*) on the basal joint of each limb. These *coxopoditic setæ*, no doubt, serve to prevent the intrusion of parasites and other foreign matters into the branchial chamber. From the mode of attachment of the six branchiæ it is obvious that they must share in the movements of the basal joints of the legs; and that, when the crayfish walks, they must be more or less agitated in the branchial chamber.

The eighteenth branchia resembles one of the eleven arthrobranchiæ in structure; but it is larger, and it is attached neither to the basal joint of the hindermost ambulatory limb, nor to its interarticular membrane, but to the sides of the thorax, above the joint. From this mode of attachment it is distinguished from the others as a *pleurobranchia* (fig. 4, *plb. 14*).

Finally, in front of this, fixed also to the walls of the thorax, above each of the two preceding pairs of ambulatory limbs, there is a delicate filament, about a sixteenth of an inch long, which has the structure of a branchial filament, and is, in fact, a rudimentary pleurobranchia (fig. 4, *plb. 12*, *plb. 13*).

The quantity of water which occupies the space left in the branchial chamber by the gills is but small, and as the respiratory surface offered by the gills is relatively very large, the air contained in this water must be rapidly exhausted, even when the crayfish is quiescent; while, when any muscular exertion takes place, the quantity of carbonic acid formed, and the demand for fresh oxygen, is at once greatly increased. For the efficient performance of the function of respiration, therefore, the water in the branchial chamber must be rapidly renewed, and there must be some arrangement by which the supply of fresh water may be proportioned to the demand. In many animals, the respiratory surface is covered with rapidly vibrating filaments, or *cilia*, by means of which a current of water is kept continually flowing over the gills, but there are none of these in the crayfish. The same object is attained, however, in another way. The anterior boundary of the branchial chamber corresponds with the cervical groove, which, as has been seen, curves downwards and then forwards, until it terminates at the sides of the space occupied by the jaws. If the branchiostegite is cut away along the groove, it will be found that it is attached to the sides of the head, which project a little beyond the anterior part of the thorax, so that there is a depression behind the sides of the head—just as there is a depression, behind a man's jaw, at the sides of the neck. Between this depression in front, the walls of the thorax internally, the branchiostegite externally, and the bases of the forceps and external foot-jaws below, a curved canal is included, by which the branchial cavity opens forwards as by a funnel. Attached to the base of the second maxilla there is a wide curved plate (fig. 4, *6*) which fits against the projection of the head, as a shirt collar might do, to carry out our previous comparison; and this scoop-shaped plate (termed the *scaphognathite*), which is concave forwards and convex backwards, can be readily moved backwards and forwards.

If a living crayfish is taken out of the water, it will be found that, as the water drains away from the branchial cavity, bubbles of air are forced out of its anterior opening.

Again, if, when a crayfish is resting quietly in the water, a little coloured fluid is allowed to run down towards the posterior opening of the branchial chamber, it will very soon be driven out from the anterior aperture, with considerable force, in a long stream. In fact, as the scaphognathite vibrates not less than three or four times in a second, the water in the funnel-shaped front passage of the branchial cavity is incessantly baled out; and, as fresh water flows in from behind to make up the loss, a current is kept constantly flowing over the gills. The rapidity of this current naturally depends on the more or less quick repetition of the strokes of the scaphognathite; and hence, the activity of the respiratory function can be accurately adjusted to the wants of the economy. Slow working of the scaphognathite answers to ordinary breathing in ourselves, quick working to panting.

A further self-adjustment of the respiratory apparatus is gained by the attachment of the six gills to the basal joints of the legs. For, when the animal exerts its muscles in walking, these gills are agitated, and thus not only bring their own surfaces more largely in contact with the water, but produce the same effect upon the other gills.

The constant oxidation which goes on in all parts of the body not only gives rise to carbonic acid; but, so far as it affects the proteinaceous constituents, it produces compounds which contain nitrogen, and these, like other waste products, must be eliminated. In the higher animals, such waste products take the form of urea, uric acid, hippuric acid, and the like; and are got rid of by the kidneys. We may, therefore, expect to find some organ which plays the part of a kidney in the crayfish; but the position of the structure, which there is much reason for regarding as the representative of the kidney, is so singular that very different interpretations have been put upon it.

On the basal joint of each antenna it is easy to see a small conical eminence with an opening on the inner side of its summit (fig. 18). The aperture (*x*) leads by a short canal into a spacious sac, with extremely delicate walls (*s*), which is lodged in the front part of the head, in front of and below the cardiac division of the stomach (*cs*). Beneath this, in a sort of recess, which corresponds with the epistoma, and with the base of the antenna, there is a discoidal body of a dull green colour, in shape somewhat like one of the fruits of the mallow, which is known as the *green gland* (*gg*). The sac narrows below like a wide funnel, and the edges of its small end are continuous with the walls of the green gland; they surround an aperture which leads into the interior of the latter structure, and conveys its products into the sac, whence they are excreted by the opening in the antennary papilla. The green gland is said to contain a substance termed *guanin* (so named because it is found in the *guano* which is the accumulated excrement of birds), a nitrogenous body analogous in some respects to uric acid, but less highly oxidated;

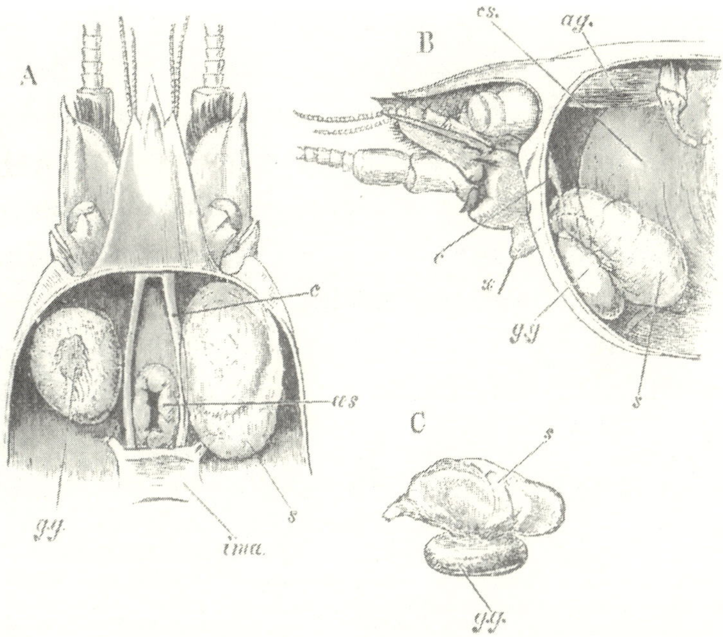

FIG. 18.—*Astacus fluviatilis.*—A, the anterior part of the body, with the dorsal portion of the carapace removed to show the position of the green glands; B, the same, with the left side of the carapace removed; C, the green gland removed from the body (all × 2). *ag*, left anterior gastric muscle; *c*, circumœsophageal commissures; *cs*, cardiac portion of stomach; *gg*. green gland, exposed in A on the left side by the removal of its sac; *ima*, intermaxillary or cephalic apodeme; *œs*, œsophagus seen in transverse section in A, the stomach being removed; *s*, sac of green gland; *x*, bristle passed from the aperture in the basal joint of the antenna into the sac.

and if this be the case, there can be little doubt that the green gland represents the kidney, and its secretion the urinary fluid, while the sac is a sort of urinary bladder.

Restricting our attention to the phenomena which have now been described, and to a short period in the life of the crayfish, the body of the animal may be regarded as a factory, provided with various pieces of machinery, by means of which certain nitrogenous and other matters are extracted from the animal and vegetable substances which serve for food, are oxidated, and are then delivered out of the factory in the shape of carbonic acid gas, guanin, and probably some other products, with which we are at present unacquainted. And there is no doubt, that if the total amount of products given out could be accurately weighed against the total amount of materials taken in, the weight of the two would be found to be identical. To put the matter in its most general shape, the body of the crayfish is a sort of focus to which certain material particles converge, in which they move for a time, and from which they are afterwards expelled in new combinations. The parallel between a whirlpool in a stream and a living being, which has often been drawn, is as just as it is striking. The whirlpool is permanent, but the particles of water which constitute it are incessantly changing. Those which enter it, on the one side, are whirled around and temporarily constitute a part of its individuality; and as they leave it on the other side, their places are made good by new comers.

Those who have seen the wonderful whirlpool, three miles below the Falls of Niagara, will not have forgotten the heaped-up wave which tumbles and tosses, a very embodiment of restless energy, where the swift stream hurrying from the Falls is compelled to make a sudden turn towards Lake Ontario. However changeful in the contour of its crest, this wave has been visible, approximately in the same place, and with the same general form, for centuries past. Seen from a mile off, it would appear to be a stationary hillock of water. Viewed closely, it is a typical expression of the conflicting impulses generated by a swift rush of material particles.

Now, with all our appliances, we cannot get within a good many miles, so to speak, of the crayfish. If we could, we should see that it was nothing but the constant form of a similar turmoil of material molecules which are constantly flowing into the animal on the one side, and streaming out on the other.

The chemical changes which take place in the body of the crayfish, are doubtless, like other chemical changes, accompanied by the evolution of heat. But the amount of heat thus generated is so small and, in consequence of the conditions under which the crayfish lives, it is so easily carried away, that it is practically insensible. The crayfish has approximately the temperature of the surrounding medium, and it is, therefore, reckoned among the cold-blooded animals.

If our investigation of the results of the process of

alimentation in a well-fed Crayfish were extended over a longer time, say a year or two, we should find that the products given out were no longer equal to the materials taken in, and the balance would be found in the increase of the animal's weight. If we inquired how the balance was distributed, we should find it partly in store, chiefly in the shape of fat; while, in part, it had been spent in increasing the plant and in enlarging the factory. That is to say, it would have supplied the material for the animal's growth. And this is one of the most remarkable respects in which the living factory differs from those which we construct. It not only enlarges itself, but, as we have seen, it is capable of executing its own repairs to a very considerable extent.

CHAPTER III.

THE PHYSIOLOGY OF THE CRAYFISH—THE MECHANISM BY WHICH THE LIVING ORGANISM ADJUSTS ITSELF TO SURROUNDING CONDITIONS AND REPRODUCES ITSELF.

IF the hand is brought near a vigorous crayfish, free to move in a large vessel of water, it will generally give a vigorous flap with its tail, and dart backwards out of reach; but if a piece of meat is gently lowered into the vessel, the crayfish will sooner or later approach and devour it.

If we ask why the crayfish behaves in this fashion, every one has an answer ready. In the first case, it is said that the animal is aware of danger, and therefore hastens away; in the second, that it knows that meat is good to eat, and therefore walks towards it and makes a meal. And nothing can seem to be simpler or more satisfactory than these replies, until we attempt to conceive clearly what they mean; and, then, the explanation, however simple it may be admitted to be, hardly retains its satisfactory character.

For example, when we say that the crayfish is "aware of danger," or "knows that meat is good to eat," what do we mean by "being aware" and "knowing"? Certainly it cannot be meant that the crayfish says to himself, as we do, "This is dangerous," "That is nice;" for the crayfish, being devoid of language, has nothing to say either to himself or any one else. And if the crayfish has not language enough to construct a proposition, it is obviously out of the question that his actions should be guided by a logical reasoning process, such as that by which a man would justify similar actions. The crayfish assuredly does not first frame the syllogism, "Dangerous things are to be avoided; that hand is dangerous; therefore it is to be avoided;" and then act upon the conclusion thus logically drawn.

But it may be said that children, before they acquire the use of language, and we ourselves, long after we are familiar with conscious reasoning, perform a great variety of perfectly rational acts unconsciously. A child grasps at a sweetmeat, or cowers before a threatening gesture, before it can speak; and any one of us would start back from a chasm opening at our feet, or stoop to pick up a jewel from the ground, "without thinking about it." And, no doubt, if the crayfish has any mind at all, his mental operations must more or less resemble those which the human mind performs without giving them a spoken or unspoken verbal embodiment.

If we analyse these, we shall find that, in many cases, distinctly felt sensations are followed by a distinct desire to perform some act, which act is accordingly performed;

while, in other cases, the act follows the sensation without one being aware of any other mental process; and, in yet others, there is no consciousness even of the sensation. As I wrote these last words, for example, I had not the slightest consciousness of any sensation of holding or guiding the pen, although my fingers were causing that instrument to perform exceedingly complicated movements. Moreover, experiments upon animals have proved that consciousness is wholly unnecessary to the carrying out of many of those combined movements by which the body is adjusted to varying external conditions.

Under these circumstances, it is really quite an open question whether a crayfish has a mind or not; moreover, the problem is an absolutely insoluble one, inasmuch as nothing short of being a crayfish would give us positive assurance that such an animal possesses consciousness; and, finally, supposing the crayfish has a mind, that fact does not explain its acts, but only shows that, in the course of their accomplishment, they are accompanied by phenomena similar to those of which we are aware in ourselves, under like circumstances.

So we may as well leave this question of the crayfish's mind on one side for the present, and turn to a more profitable investigation, namely, that of the order and connexion of the physical phenomena which intervene between something which happens in the neighbourhood of the animal and that other something which responds to it, as an act of the crayfish.

Whatever else it may be, this animal, so far as it is acted upon by bodies around it and reacts on them, is a piece of mechanism, the internal works of which give rise to certain movements when it is affected by particular external conditions; and they do this in virtue of their physical properties and connexions.

Every movement of the body, or of any organ of the body, is an effect of one and the same cause, namely, muscular contraction. Whether the crayfish swims or walks, or moves its antennæ, or seizes its prey, the immediate cause of the movements of the parts which bring about, or constitute, these bodily motions is to be sought in a change which takes place in the flesh, or muscle, which is attached to them. The change of place which constitutes any movement is an effect of a previous change in the disposition of the molecules of one or more muscles; while the direction of that movement depends on the connexions of the parts of the skeleton with one another, and of the muscles with them.

The muscle of the crayfish is a dense, white substance; and if a small portion of it is subjected to examination it will be found to be very easily broken up into more or less parallel bundles of fine fibres. Each of these fibres is generally found to be ensheathed in a fine transparent membrane, which is called the *sarcolemma*, within which is contained the proper substance of the muscle. When quite fresh and living, this substance is soft and semi-fluid, but it hardens and becomes solid immediately after death.

Examined, with high magnifying powers, in this

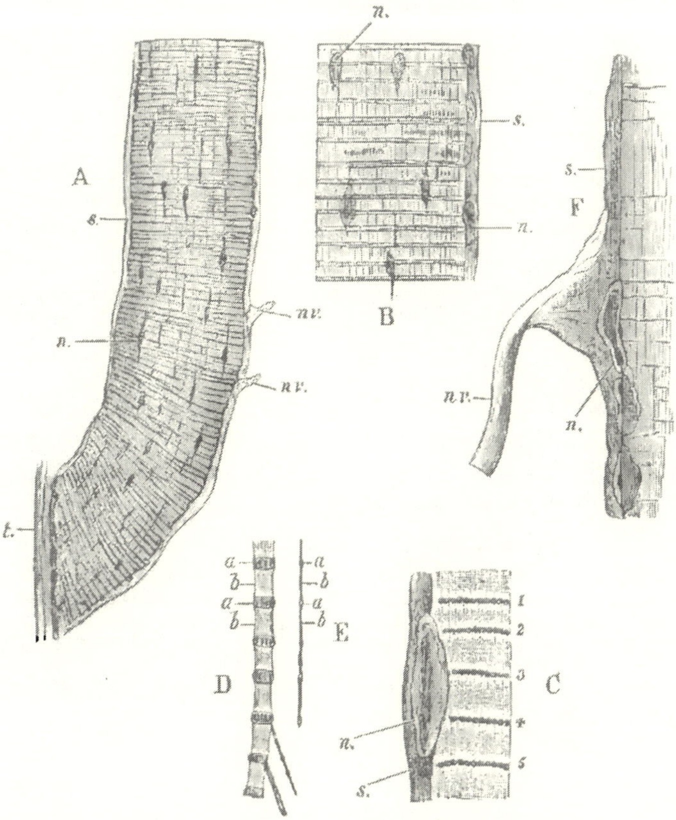

FIG. 19.—*Astacus fluviatilis.*—A, a single muscular fibre; transverse diameter $\frac{1}{110}$th of an inch; B, a portion of the same more highly magnified; C, a smaller portion still more highly magnified; D and E, the splitting up of a part of fibre into fibrillæ; F, the connexion of a nervous with a muscular fibre which has been treated with acetic acid. *a*, darker, and *b*, clearer portions of the fibrillæ; *n*, nucleus of sarcolemma; *nv*, nerve fibre; *s*, sarcolemma; *t*, tendon; 1—5, successive dark bands answering to the darker portions, *a*, of each fibrilla.

condition, the muscle-substance appears marked by very regular transverse bands, which are alternately opaque and transparent; and it is characteristic of the group of animals to which the crayfish belongs that their muscle-substance has this striped character in all parts of the body.

A greater or less number of these fibres, united into one or more bundles, constitutes a muscle; and, except when these muscles surround a cavity, they are fixed at each end to the hard parts of the skeleton. The attachment is frequently effected by the intermediation of a dense, fibrous, often chitinous, substance, which constitutes the *tendon* (fig. 19, A; *t*) of the muscle.

The property of the living muscle, which enables it to be the cause of motion, is this: Every muscular fibre is capable of suddenly changing its dimensions, in such a manner that it shortens and becomes proportionately thicker. Hence the absolute bulk of the fibre remains practically unchanged. From this circumstance, muscular *contraction*, as the change of form of a muscle is called, is radically different from the process which commonly goes by the same name in other things, and which involves a diminution of bulk.

The contraction of muscle takes place with great force, and, of course, if the parts to which its ends are fixed are both free to move, they are brought nearer at the moment of contraction: if one only is free to move that is approximated to the fixed part; and if the muscular fibre surrounds a cavity, the cavity is lessened when the muscle contracts. This is the whole source of motor power in the crayfish machine. The results produced by the exertion of that power depend upon the manner

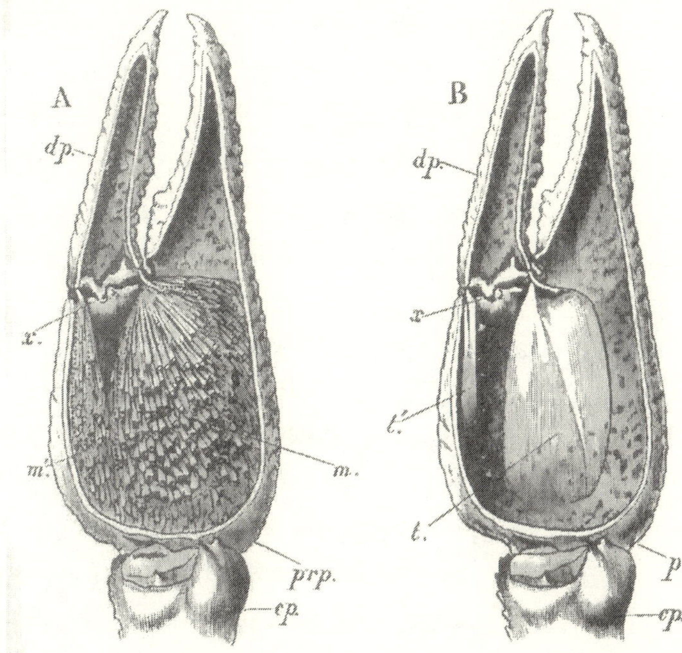

FIG. 20.—*Astacus fluviatilis.*—The chela of the forceps, with one side cut away to show, in A, the muscles, in B, the tendons (× 2). *cp*, carpopodite ; *prp*, propodite ; *dp*, dactylopodite ; *m*, adductor muscle ; *m'*, abductor muscle ; *t*, tendon of adductor muscle ; *t'*, tendon of abductor muscle ; *x*, hinge.

in which the parts to which the muscles are attached are connected with one another.

One example of this has already been given in the curious mechanism of the gastric mill. Another may be found in the chela which terminates the forceps. If the articulation of the last joint (fig. 20, *dp*) with the one which precedes it (*prp*) is examined, it will be found that the base of the terminal segment (*dp*) turns on two hinges (*x*), formed by the hard exoskeleton and situated at opposite points of the diameter of the base, on the penultimate segment; and these hinges are so disposed that the last joint can be moved only in one plane, to or from the produced angle of the penultimate segment (*prp*), which forms the fixed claw of the chela. Between the hinges, on both the inner and the outer sides of the articulation, the exoskeleton is soft and flexible, and allows the terminal segment to play easily through a certain arc. It is by this arrangement that the direction and the extent of the motion of the free claw of the chela are determined. The source of the motion lies in the muscles which occupy the interior of the enlarged penultimate segment of the limb. Two muscles, one of very great size (*m*), the other smaller (*m'*), are fastened by one end to the exoskeleton of this segment. The fibres of the larger muscle converge to be fixed into the two sides of a long flat process of the chitinous cuticula, on the inner side of the base of the terminal segment, which serves as a tendon (*t*); while those of the smaller muscle are similarly attached to a like process which proceeds from the outer side of the base of the terminal segment (*t'*). It is obvious that, when the latter muscle shortens it must move the apex of the terminal segment (*dp*) away from the end of the fixed claw; while, when the former contracts, the end of the terminal segment is brought towards that of the fixed claw.

A living crayfish is able to perform very varied movements with its pincers. When it swims backwards, these limbs are stretched straight out, parallel with one another, in front of the head; when it walks, they are usually carried like arms bent at the elbow, the "forearm" partly resting on the ground; on being irritated, the crayfish sweeps the pincers round in any direction to grasp the offending body; when prey is seized, it is at once conveyed, with a circular motion, towards the region of the mouth. Nevertheless, these very varied actions are all brought about by a combination of simple flexions and extensions, each of which is effected in the exact order, and to the exact extent, needful to bring the chela into the position required.

The skeleton of the stem of the limb which bears the chela is, in fact, divided into four moveable segments; and each of these is articulated with the segments on each side of it by a hinge of just the same character as that which connects the moveable claw of the chela with the penultimate segment, while the basal segment is similarly articulated with the thorax.

If the axes of all these articulations * were parallel, it is obvious that, though the limb might be moved as a whole through a considerable arc, and might be bent in various

* By axis of the articulation is meant a line drawn through the pair of hinges which constitute it.

degrees, yet all its movements would be limited to one plane. But, in fact, the axes of the successive articulations are nearly at right angles to one another; so that, if the segments are successively either extended or flexed, the chela describes a very complicated curve; and by varying the extent of flexion or extension of each segment, this curve is susceptible of endless variation. It would probably puzzle a good mathematician to say exactly what position should be given to each segment, in order to bring the chela from any given position into any other; but if a lively crayfish is incautiously seized, the experimenter will find, to his cost, that the animal solves the problem both rapidly and accurately.

The mechanism by which the retrograde swimming of the crayfish is effected, is no less easily analysed. The apparatus of motion is, as we have seen, the abdomen, with its terminal five-pointed flapper. The rings of the abdomen are articulated together by joints (fig. 21, ×) situated a little below the middle of the height of the rings, at opposite ends of transverse lines, at right angles to the long axis of the abdomen.

Each ring consists of a dorsal, arched portion, called the *tergum* (fig. 21; fig. 36, p. 142, *t. XIX*), and a nearly flat ventral portion, which is the *sternum* (fig. 36, *st. XIX*). Where these two join, a broad plate is sent down on each side, which overlaps the bases of the abdominal appendages, and is known as the *pleuron* (fig. 36, *pl. XIX*).

The sterna are all very narrow, and are connected together by wide spaces of flexible exoskeleton.

When the abdomen is made straight, it will be found that these *intersternal* membranes are stretched as far as they will yield. On the other hand, when the abdomen

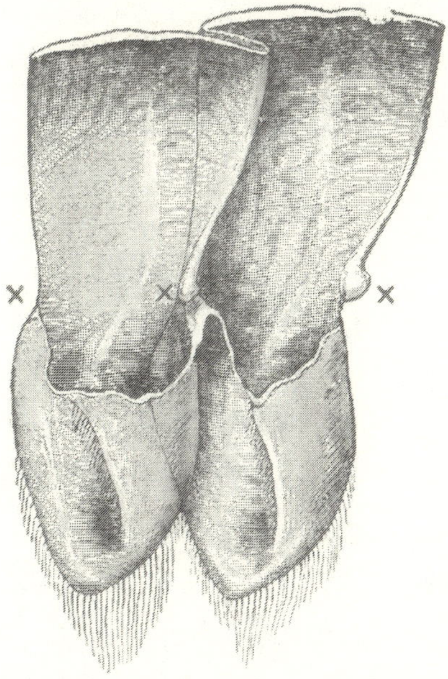

FIG. 21.—*Astacus fluviatilis*.—Two of the abdominal somites, in vertical section, seen from the inner side, to show ×, ×, the hinges by which they are articulated with one another (× 3). The anterior of the two somites is that to the right of the figure.

is bent up as far as it will go, the sterna come close together, and the intersternal membranes are folded.

The terga are very broad; so broad, in fact, that each overlaps its successor, when the abdomen is straightened or extended, for nearly half its length in the middle line; and the overlapped surface is smooth, convex, and marked off by a transverse groove from the rest of the tergum as an *articular facet.* The front edge of the articular facet is continued into a sheet of flexible cuticula, which turns back, and passes as a loose fold to the hinder edge of the overlapping tergum (fig. 21). This tergal *interarticular membrane* allows the terga to move as far as they can go in flexion; whilst, in extreme extension, they are but slightly stretched. But, even if the intersternal membranes presented no obstacle to excessive extension of the abdomen, the posterior free edge of each tergum fits into the groove behind the facet in the next in such a manner, that the abdomen cannot be made more than very slightly concave upwards without breaking.

Thus the limits of motion of the abdomen, in the vertical direction, are from the position in which it is straight, or has even a very slight upward concavity, to that in which it is completely bent upon itself, the telson being brought under the bases of the hinder thoracic limbs. No lateral movement between the somites of the abdomen is possible in any of its positions. For, when it is straight, lateral movement is hindered not only by the extensive overlapping of the terga, but also by the manner in which the hinder edges of the pleura of each of the four middle somites overlap the front edges of their successors. The pleura of the second somite are much larger than any of the others, and their front edges overlap the small pleura of the first abdominal somite; and when the abdomen is much flexed, these pleura even ride over the posterior edges of the branchiostegites. In the position of extension, the overlap of the terga is great, while that of the pleura of the middle somites is small. As the abdomen passes from extension to flexion, the overlap of the terga of course diminishes; but any decrease of resistance to lateral strains which may thus arise, is compensated by the increasing overlap of the pleura, which reaches its maximum when the abdomen is completely flexed.

It is obvious that longitudinal muscular fibres fixed into the exoskeleton, above the axes of the joints, must bring the centres of the terga of the somites closer together, when they contract; while muscular fibres attached below the axes of the joints must approximate the sterna. Hence, the former will give rise to extension, and the latter to flexion, of the abdomen as a whole.

Now there are two pairs of very considerable muscles disposed in this manner. The dorsal pair, or the *extensors* of the abdomen (fig. 22, *e.m*), are attached in front to the side walls of the thorax, thence pass backwards into the abdomen, and divide into bundles, which are fixed to the inner surfaces of the terga of all the somites. The other pair, or the *flexors* of the abdomen (*f.m*) constitute a very much larger mass of muscle, the fibres of which are curiously twisted, like the strands of a rope. The front end of this double rope is fixed to a series of processes of the exoskeleton of the thorax, called *apodemata*, some of which roof over the sternal blood-sinuses and the thoracic part of the nervous system; while, in the abdomen, its strands are attached to the sternal exoskeleton of all the somites and extend, on each side of the rectum, to the telson.

When the exoskeleton is cleaned by maceration, the

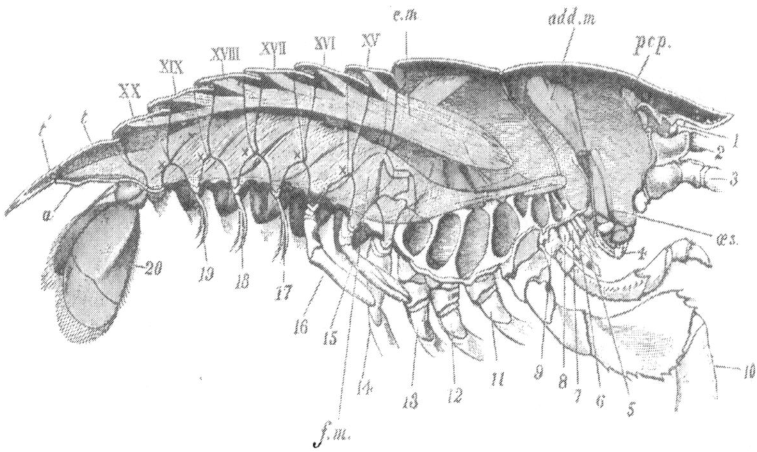

FIG. 22.—*Astacus fluviatilis.*—A longitudinal section of the body to show the principal muscles and their relations to the exoskeleton (nat. size). *a*, the vent; *add.m*, adductor muscle of mandible; *e.m*, extensor, and *f.m*, flexor muscle of abdomen; *œs*, œsophagus; *pcp*, procephalic process; *t,t'*, the two segments of the telson; *XV—XX*, the abdominal somites; *1—20*, the appendages; ×, ×, hinges between the successive abdominal somites.

abdomen has a slight curve, dependent upon the form and the degree of elasticity possessed by its different parts; and, in a living crayfish at rest, it will be observed that the curvature of the abdomen is still more marked. Hence it is ready either for extension or for flexion.

A sudden contraction of the flexor muscles instantly increases the ventral curvature of the abdomen, and throws the tail fin, the two side lobes of which are spread out, forwards; while the body is propelled backwards by the reaction of the water against the stroke. Then the flexor muscles being relaxed, the extensor muscles come into play; the abdomen is straightened, but less violently and with a far weaker stroke on the water, in consequence of the less strength of the extensors and of the folding up of the lateral plates of the fin, until it comes into the position requisite to give full force to a new downward and forward stroke. The tendency of the extension of the abdomen is to drive the body forward; but from the comparative weakness and the obliquity of its stroke, its practical effect is little more than to check the backward motion conferred upon the body by the flexion of the abdomen.

Thus, every action of the crayfish, which involves motion, means the contraction of one or more muscles. But what sets muscle contracting? A muscle freshly removed from the body may be made to contract in various ways, as by mechanical or chemical irritation, or by an electrical shock; but, under natural conditions, there is only one cause of muscular contraction, and that is the activity of a nerve. Every muscle is supplied with one or more nerves. These are delicate threads which, on microscopic examination, prove to be bundles of fine tubular filaments, filled with an apparently structureless substance of gelatinous consistency, the *nerve fibres* (fig. 23). The nerve bundle which passes to a muscle breaks up into smaller bundles and, finally, into separate fibres, each of which ultimately terminates by becoming continuous with the substance of a muscular fibre fig. 19, F.) Now the peculiarity of a muscle nerve, or *motor* nerve, as it is called, is that irritation of the nerve fibre at any part of its length, however distant from the muscle,

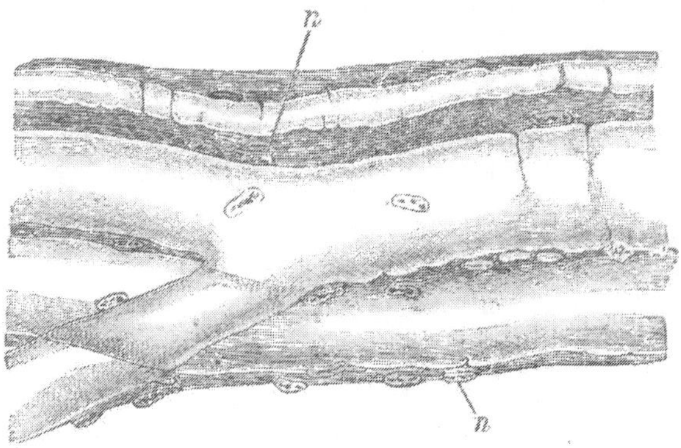

FIG. 23.—*Astacus fluviatilis.*—Three nerve fibres, with the connective tissue in which they are imbedded. (Magnified about 250 diameters.) *n*, nuclei.

brings about muscular contraction, just as if the muscle itself were irritated. A change is produced in the molecular condition of the nerve at the point of irritation; and this change is propagated along the nerve, until it reaches the muscle, in which it gives rise to that change in the arrangement of its molecules, the most obvious effect of which is the sudden alteration of form which we call muscular contraction.

If we follow the course of the motor nerves in a

direction away from the muscles to which they are distributed, they will be found, sooner or later, to terminate in *ganglia* (fig. 24 A. *gl.c*; fig. 25, *gn. 1—13.*) A ganglion is a body which is in great measure composed of

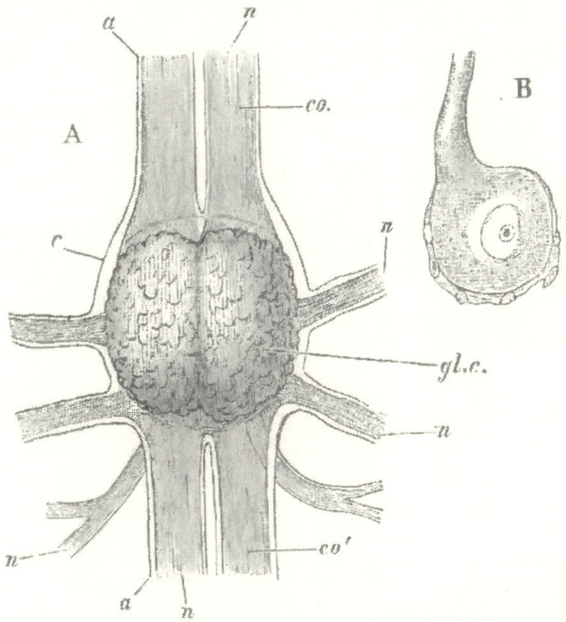

FIG. 24.—*Astacus fluviatilis.*—A, one of the (double) abdominal ganglia, with the nerves connected with it (× 25); B, a nerve cell or ganglionic corpuscle (× 250). *a*, sheath of the nerves; *c*, sheath of the ganglion; *co*, *co'*, commissural cords connecting the ganglia with those in front, and those behind them. *gl.c.* points to the ganglionic corpuscles of the ganglia; *n*, nerve fibres.

nerve fibres; but, interspersed among these, or disposed around them, there are peculiar structures, which are termed *ganglionic corpuscles*, or *nerve cells* (fig. 24, B.) These are nucleated cells, not unlike the epithelial cells which have been already mentioned, but which are larger

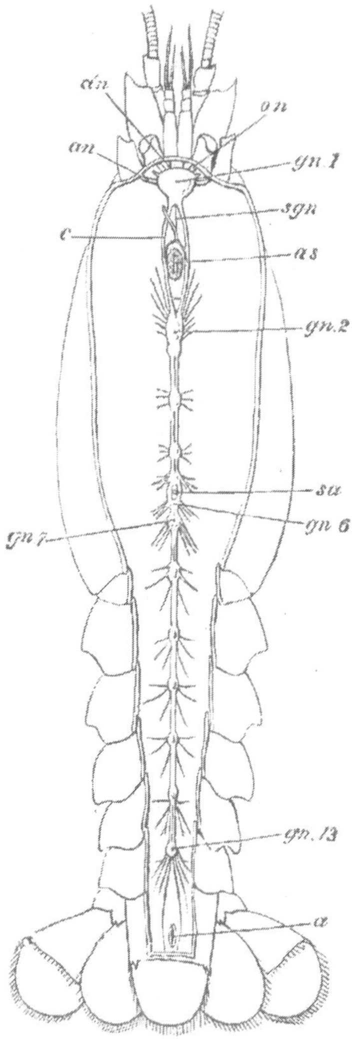

FIG. 25.—*Astacus fluviatilis.*—The central nervous system seen from above (nat. size). *a*, vent ; *an*, antennary nerve ; *a'n*, antennulary nerve ; *c*, circumœsophageal commissures ; *gn*. 1, supraœsophageal ganglion ; *gn*. 2, infraœsophageal ganglion ; *gn*. 6, fifth thoracic ganglion ; *gn*. 7, last thoracic ganglion ; *gn*. 13, last abdominal ganglion ; *œs*, œsophagus in cross section ; *on*, optic nerve; *sa*, sternal artery in cross section ; *sgn*, stomatogastric nerve.

and often give off one or more processes. These processes, under favourable circumstances, can be traced into continuity with nerve fibres.

The chief ganglia of the crayfish are disposed in a longitudinal series in the middle line of the ventral aspect of the body close to the integument (fig. 25). In the abdomen, for example, six ganglionic masses are readily observed, one lying over the sternum of each somite, connected by longitudinal bands of nerve fibres, and giving off branches to the muscles. On careful examination, the longitudinal connecting bands, or *commissures* (fig. 24, *co*), are seen to be double, and each mass appears slightly bilobed. In the thorax, there are six, larger, double ganglionic masses, likewise connected by double commissures; and the most anterior of these, which is the largest (fig. 25, *gn*. 2), is marked at the sides by notches, as if it were made up of several pairs of ganglia, run together into one continuous whole. In front of this, two commissures (*c*) pass forwards, separating widely, to give room for the gullet (*œs*), which passes between them; while in front of the gullet, just behind the eyes, they unite with a transversely elongated mass of ganglionic substance (*gn*. 1), termed the *brain*, or *cerebral ganglion*.

All the motor nerves, as has been said, are traceable, directly or indirectly, to one or other of these thirteen sets of ganglia; but other nerves are given off from the ganglia, which cannot be followed into any muscle. In fact, these nerves go either to the integument or to the organs of sense, and they are termed *sensory nerves.*

When a muscle is connected by its motor nerve with a ganglion, irritation of that ganglion will bring about the contraction of the muscle, as well as if the motor nerve itself were irritated. Not only so; but if a sensory nerve, which is in connexion with the ganglion, is irritated, the same effect is produced; moreover, the sensory nerve itself need not be excited, but the same result will take place, if the organ to which it is distributed is stimulated. Thus the nervous system is fundamentally an apparatus by which two separate, and it may be distant, parts of the body, are brought into relation with one another; and this relation is of such a nature, that a change of state arising in the one part is followed by the propagation of changes along the sensory nerve to the ganglion, and from the ganglion to the other part; where, if that part happens to be muscle, it produces contraction. If one end of a rod of wood, twenty feet long, is applied to a sounding-board, the sound of a tuning-fork held against the opposite extremity will be very plainly heard. Nothing can be seen to happen in the wood, and yet its molecules are certainly set vibrating, at the same rate as the tuning-fork vibrates; and when, after travelling rapidly along the wood, these vibrations affect the sounding-board, they give rise to vibrations of the molecules of the air, which reaching the ear, are converted into an audible note. So in the nerve tract:

no apparent change is effected in it by the irritation at one end; but the rate at which the molecular change produced travels can be measured; and, when it reaches the muscle, its effect becomes visible in the change of form of the muscle. The molecular change would take place just as much if there were no muscle connected with the nerve, but it would be no more apparent to ordinary observation than the sound of the tuning-fork is audible in the absence of the sounding-board.

If the nervous system were a mere bundle of nerve fibres extending between sensory organs and muscles, every muscular contraction would require the stimulation of that special point of the surface on which the appropriate sensory nerve ended. The contraction of several muscles at the same time, that is, the combination of movements towards one end, would be possible only if the appropriate nerves were severally stimulated in the proper order, and every movement would be the direct result of external changes. The organism would be like a piano, which may be made to give out the most complicated harmonies, but is dependent for their production on the depression of a separate key for every note that is sounded. But it is obvious that the crayfish needs no such separate impulses for the performance of highly complicated actions. The simple impression made on the organs of sensation in the two examples with which we started, gives rise to a train of complicated and accurately co-ordinated muscular contractions. To carry the analogy of the musical instrument further, striking a single key gives rise, not to a single note, but to a more or less elaborate tune; as if the hammer struck not a single string, but pressed down the stop of a musical box.

It is in the ganglia that we must look for the analogue of the musical box. A single impulse conveyed by a sensory nerve to a ganglion, may give rise to a single muscular contraction, but more commonly it originates a series of such, combined to a definite end.

The effect which results from the propagation of an impulse along a nerve fibre to a ganglionic centre, whence it is, as it were, reflected along another nerve fibre to a muscle, is what is termed a *reflex action*. As it is by no means necessary that sensation should be a concomitant of the first impulse, it is better to term the nerve fibre which carries it *afferent* rather than sensory; and, as other phenomena besides those of molar motion may be the ultimate result of the reflex action, it is better to term the nerve fibre which transmits the reflected impulse *efferent* rather than motor.

If the nervous commissures between the last thoracic and the first abdominal ganglia are cut, or if the thoracic ganglia are destroyed, the crayfish is no longer able to control the movements of the abdomen. If the forepart of the body is irritated, for example, the animal makes no effort to escape by swimming backwards. Nevertheless, the abdomen is not paralysed, for, if it be irritated, it will flap vigorously. This is a case of pure reflex action. The stimulus is conveyed to the abdominal ganglia through afferent nerves, and is reflected from them, by efferent nerves, to the abdominal muscles.

But this is not all. Under these circumstances it will be seen that the abdominal limbs all swing backwards and forwards, simultaneously, with an even stroke; while the vent opens and shuts with a regular rhythm. Of course, these movements imply correspondingly regular alternate contractions and relaxations of certain sets of muscles; and these, again, imply regularly recurring efferent impulses from the abdominal ganglia. The fact that these impulses proceed from the abdominal ganglia, may be shown in two ways: first, by destroying these ganglia in one somite after another, when the movements in each somite at once permanently cease; and, secondly, by irritating the surface of the abdomen, when the movements are temporarily inhibited by the stimulation of the afferent nerves. Whether these movements are properly reflex, that is, arise from incessant new afferent impulses of unknown origin, or whether they depend on the periodical accumulation and discharge of nervous energy in the ganglia themselves, or upon periodical exhaustion and restoration of the irritability of the muscles, is unknown. It is sufficient for the present purpose to use the facts as evidence of the peculiar co-ordinative function of ganglia.

The crayfish, as we have seen, avoids light; and the slightest touch of one of its antennæ gives rise to active motions of the whole body. In fact, the animal's position and movements are largely determined by the influences received through the feelers and the eyes. These receive their nerves from the cerebral ganglia; and, as might be expected, when these ganglia are extirpated, the crayfish exhibits no tendency to get away from the light, and the feelers may not only be touched, but sharply pinched, without effect. Clearly, therefore, the cerebral ganglia serve as a ganglionic centre, by which the afferent impulses derived from the feelers and the eyes are transmuted into efferent impulses. Another very curious result follows upon the extirpation of the cerebral ganglia. If an uninjured crayfish is placed upon its back, it makes unceasing and well-directed efforts to turn over; and if everything else fails, it will give a powerful flap with the abdomen, and trust to the chapter of accidents to turn over as it darts back. But the brainless crayfish behaves in a very different way. Its limbs are in incessant motion, but they are "all abroad;" and if it turns over on one side, it does not seem able to steady itself, but rolls on to its back again.

If anything is put between the chelæ of an uninjured crayfish, while on its back, it either rejects the object at once, or tries to make use of it for leverage to turn over. In the brainless crayfish a similar operation gives rise to a very curious spectacle.* If the object, whatever it be

* My attention was first drawn to these phenomena by my friend Dr. M. Foster, F.R.S., to whom I had suggested the desirableness of an experimental study of the nerve physiology of the crayfish.

—a bit of metal, or wood, or paper, or one of the animal's own antennæ—is placed between the chelæ of the forceps, it is at once seized by them, and carried backwards; the chelate ambulatory limbs are at the same time advanced, the object seized is transferred to them, and they at once tuck it between the external maxillipedes, which, with the other jaws, begin vigorously to masticate it. Sometimes the morsel is swallowed; sometimes it passes out between the anterior jaws, as if deglutition were difficult. It is very singular to observe that, if the morsel which is being conveyed to the mouth by one of the forceps is pulled back, the forceps and the chelate ambulatory limbs of the other side are at once brought forward to secure it. The movements of the limbs are, in short, adjusted to meet the increased resistance.

All these phenomena cease at once, if the thoracic ganglia are destroyed. It is in these, therefore, that the simple stimulus set up by the contact of a body with, for example, one of the forceps, is translated into all the surprisingly complex and accurately co-ordinated movements, which have been described. Thus the nervous system of the crayfish may be regarded as a system of co-ordinating mechanisms, each of which produces a certain action, or set of actions, on the receipt of an appropriate stimulus.

When the crayfish comes into the world, it possesses in its neuro-muscular apparatus certain innate potentialities of action, and will exhibit the corresponding acts, under the influence of the appropriate stimuli. A large proportion of these stimuli come from without through the organs of the senses. The greater or less readiness of each sense organ to receive impulses, of the nerves to transmit them, and of the ganglia to give rise to combined impulses, is dependent at any moment upon the physical condition of these parts; and this, again, is largely modified by the amount and the condition of the blood supplied. On the other hand, a certain number of these stimuli are doubtless originated by changes within the various organs which compose the body, including the nerve centres themselves.

When an action arises from conditions developed in the interior of an animal's body, inasmuch as we cannot perceive the antecedent phenomena, we call such an action "spontaneous;" or, when in ourselves we are aware that it is accompanied by the idea of the action, and the desire to perform it, we term the act "voluntary." But, by the use of this language, no rational person intends to express the belief that such acts are uncaused or cause themselves. "Self-causation" is a contradiction in terms; and the notion that any phenomenon comes into existence without a cause, is equivalent to a belief in chance, which one may hope is, by this time, finally exploded.

In the crayfish, at any rate, there is not the slightest reason to doubt that every action has its definite physical cause, and that what it does at any moment would be as clearly intelligible, if we only knew all the internal and external conditions of the case, as the striking of a clock is to any one who understands clockwork.

The adjustment of the body to varying external conditions, which is one of the chief results of the working of the nervous mechanism, would be far less important from a physiological point of view than it is, if only those external bodies which come into direct contact with the organism * could affect it; though very delicate influences of this kind take effect on the nervous apparatus through the integument.

It is probable that the *setæ*, or hairs, which are so generally scattered over the body and the appendages, are delicate tactile organs. They are hollow processes of the chitinous cuticle, and their cavities are continuous with narrow canals, which traverse the whole thickness of the cuticle, and are filled by a prolongation of the subjacent proper integument. As this is supplied with nerves, it is likely that fine nerve fibres reach the bases of the hairs, and are affected by anything which stirs these delicately poised levers.

* It may be said that, strictly speaking, only those external bodies which are in direct contact with the organism do affect it—as the vibrating ether, in the case of luminous bodies; the vibrating air or water, in the case of sonorous bodies; odorous particles, in the case of odorous bodies: but I have preferred the ordinary phraseology to a pedantically accurate periphrasis.

There is much reason to believe that odorous bodies affect crayfish; but it is very difficult to obtain experi-

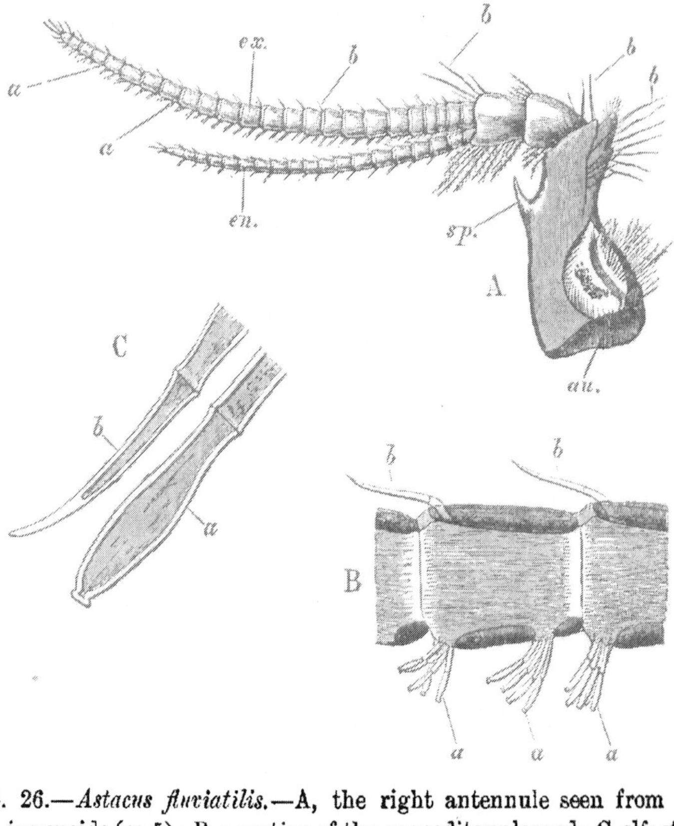

FIG. 26.—*Astacus fluviatilis.*—A, the right antennule seen from the inner side (× 5); B, a portion of the exopodite enlarged; C, olfactory appendage of the exopodite; *a*, front view; *b*, side view (× 300); *a*, olfactory appendages; *au*, auditory sac, supposed to be seen through the wall of the basal joint of the antennule; *b*, setæ; *en*, endopodite; *ex*, exopodite; *sp.* spine of the basal joint.

mental evidence of the fact. However, there is a good deal of analogical ground for the supposition that some peculiar structures, which are evidently of a sensory nature, developed on the under side of the outer branch of the antennule, play the part of an olfactory apparatus.

Both the outer (fig. 26 A. *ex*) and the inner (*en*) branches of the antennule are made up of a number of delicate ring-like segments, which bear fine setæ (*b*) of the ordinary character.

The inner branch, which is the shorter of the two, possesses only these setæ; but the under surface of each of the joints of the outer branch, from about the seventh or eighth to the last but one, is provided with two bundles of very curious appendages (fig. 27, A, B, C, *a*), one in front and one behind. These are rather more than 1-200th of an inch long, very delicate, and shaped like a spatula, with a rounded handle and a flattened somewhat curved blade, the end of which is sometimes truncated, sometimes has the form of a prominent papilla. There is a sort of joint between the handle and the blade, such as is found between the basal and the terminal parts of the ordinary setæ, with which, in fact, these processes entirely correspond in their essential structure. A soft granular tissue fills the interior of each of these problematical structures, to which Leydig, their discoverer, ascribes an olfactory function.

It is probable that the crayfish possesses something analogous to taste, and a very likely seat for the organ of this function is in the upper lip and the metastoma; but if the organ exists it possesses no structural peculiarities by which it can be identified.

There is no doubt, however, as to the special recipients of sonorous and luminous vibrations; and these are of particular importance, as they enable the nervous machinery to be affected by bodies indefinitely remote from it, and to change the place of the organism in relation to such bodies.

Sonorous vibrations are enabled to act as the stimulants of a special nerve (fig. 25, *a'n*) connected with the brain, by means of the very curious *auditory sacs* (fig. 26, A, *au*) which are lodged in the basal joints of the antennules.

Each of these joints is trihedral, the outer face being convex; the inner, applied to its fellow, flat; and the upper, on which the eyestalk rests, concave. On this upper face there is a narrow elongated oval aperture, the outer lip of which is beset with a flat brush of long close-set setæ, which lie horizontally over the aperture, and effectually close it. The aperture leads into a small sac (*au*) with delicate walls formed by a chitinous continuation of the general cuticula. The inferior and posterior wall of the sac is raised up along a curved line into a ridge which projects into its interior (fig. 27, A, *r*). Each side of this ridge is beset with a series of delicate setæ (*as*), the longest of which measures about $\frac{1}{50}$th of an inch; they thus form a longitudinal band bent upon itself. These *auditory setæ* project into the fluid contents of the sac, and their apices are for the most part imbedded in a gelatinous mass, which contains irregular particles of sand

and sometimes of other foreign matter. A nerve (*n n'*,) is distributed to the sac, and its fibres enter the bases of the hairs, and may be traced to their apices, where they end in peculiar elongated rod-like bodies (fig. 27, C). Here is an auditory organ of the simplest description.

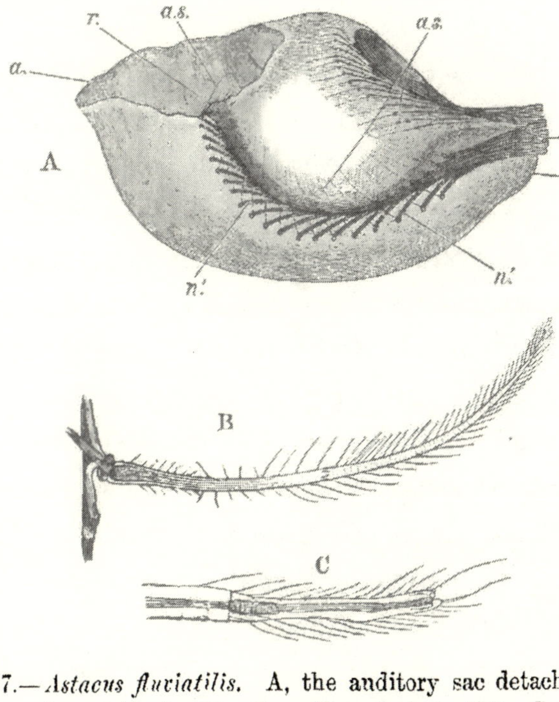

FIG. 27.—*Astacus fluviatilis.* A, the auditory sac detached and seen from the outside (×15); B, auditory hair (×100); C, the distal extremity of the same more highly magnified. *a*, aperture of sac; *as*, auditory setæ; *b*, its inner or posterior extremity; *n n'*, nerves; *r*, ridge.

It retains, in fact, throughout life, the condition of a simple sac or involution of the integument, such as is that of the vertebrate ear in its earliest stage.

The sonorous vibrations transmitted through the water in which the crayfish lives to the fluid and solid contents of the auditory sac are taken up by the delicate hairs of the ridge, and give rise to molecular changes which traverse the auditory nerves and reach the cerebral ganglia.

The vibrations of the luminiferous ether are brought to bear upon the free ends of two large bundles of nerve fibres, termed the optic nerves (fig. 25, *on*), which proceed directly from the brain, by means of a highly complex *eye*. This is an apparatus, which, in part, sorts out the rays of light into as many very small pencils as there are separate endings of the fibres of the optic nerve, and, in part, serves as the medium by which the luminous vibrations are converted into molecular nerve changes.

The free extremity of the eyestalk presents a convex, soft, and transparent surface, limited by an oval contour. The cuticle in this region, which is termed the *cornea*, (fig. 28, *a*), is, in fact, somewhat thinner and less distinctly laminated than in the rest of the eyestalk, and it contains no calcareous matter. But it is directly continuous with the rest of the exoskeleton of the eyestalk, to which it stands in somewhat the same relation as the soft integument of an articulation does to the adjacent hard parts.

The *cornea* is divided into a great number of minute, usually square facets, by faint lines, which cross it from side to side nearly at right angles with one another. A longitudinal section shows that both the horizontal and the vertical contours of the cornea are very nearly semicircular, and that the lines which mark off the facets merely arise from a slight modification of its substance between the facets. The outer contour of each facet forms part

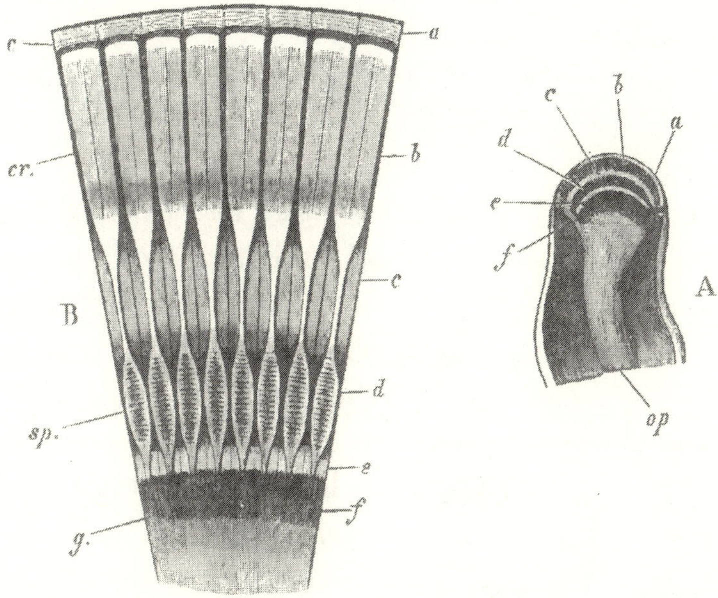

FIG. 28.—*Astacus fluviatilis.*—A, a vertical section of the eye-stalk (× 6); B, a small portion of the same, showing the visual apparatus more highly magnified; *a*, cornea; *b*, outer dark zone; *c*, outer white zone; *d*, middle dark zone; *e*, inner white zone; *f*, inner dark zone; *cr*, crystalline cones; *g*, optic ganglion; *op*, optic nerve; *sp*. striated spindles.

of the general curvature of the outer face of the cornea; the inner contour sometimes exhibits a slight deviation from the general curvature of the inner face, but usually nearly coincides with it.

When a longitudinal or a transverse section is taken through the whole eyestalk, the optic nerve (fig. 28, A, *op*) is seen to traverse its centre. At first narrow and cylindrical, it expands towards its extremity into a sort of bulb (B, *g*), the outer surface of which is curved in correspondence with the inner surface of the cornea. The terminal half of the bulb contains a great quantity of dark colouring matter or pigment, and, in section, appears as what may be termed the *inner dark zone* (*f*). Outside this, and in connection with it, follows a white line, the *inner white zone* (*e*), then comes a *middle dark zone* (*d*); outside this an outer pale band, which may be called the *outer white zone* (*c*), and between this and the cornea (*a*) is another broad band of dark pigment, the *outer dark zone* (*b*).

When viewed under a low power, by reflected light, this outer dark zone is seen to be traversed by nearly parallel straight lines, each of which starts from the boundary between two facets, and can be followed inwards through the outer white zone to the middle dark zone. Thus the whole substance of the eye between the outer surface of the bulb of the optic nerve and the inner surface of the cornea is marked out into as many segments as the cornea has facets; and each segment has the form of a wedge or slender pyramid, the base of which is four-sided, and is applied against the inner surface of one of the facets of the cornea, while its summit lies in the middle dark zone. Each of these *visual pyramids* consists of an axial structure, the *visual rod,* invested by a sheath. The latter extends inwards from the margin of each facet of the cornea, and contains pigment in two regions of its length, the intermediate space being devoid of pigment. As the position of the pigmented regions in relation to the length of the pyramid is always the same, the pigmented regions necessarily take the form of two consecutive zones when the pyramids are in their natural position.

The visual rod consists of two parts, an external *crystalline cone* (fig. 28, B, *cr*), and an internal *striated spindle* (*sp*). The *crystalline cone* consists of a transparent glassy-looking substance, which may be made to split up longitudinally into four segments. Its inner end narrows into a filament which traverses the outer white zone, and, in the middle dark zone, thickens into a four-sided spindle-shaped transparent body, which appears transversely striated. The inner end of this *striated spindle* narrows again, and becomes continuous with nerve fibres which proceed from the surface of the optic bulb.

The exact mode of cónnection of the nerve-fibres with the visual rods is not certainly made out, but it is probable that there is direct continuity of substance, and that each rod is really the termination of a nerve fibre.

Eyes having essentially the same structure as that of

the crayfish are very widely met with among *Crustacea* and *Insecta,* and are commonly known as *compound eyes.* In many of these animals, in fact, when the cornea is removed, each facet is found to act as a separate lens; and when proper arrangements are made, as many distinct pictures of external objects are found behind it as there are facets. Hence the notion suggested itself that each visual pyramid is a separate eye, similar in principle of construction to the human eye, and forming a picture of so much of the external world as comes within the range of its lens, upon a retina supposed to be spread out on the surface of the crystalline cone, as the human retina is spread over the surface of the vitreous humour.

But, in the first place, there is no evidence, nor any probability, that there is anything corresponding to a retina on the outer face of the crystalline cone; and secondly, if there were, it is incredible that, with such an arrangement of the refractive media as exists in the cornea and crystalline cones, rays proceeding from points in the external world should be brought to a focus in correspondingly related points of the surface of the supposed retina. But without this no picture could be formed, and no distinct vision could take place. It is very probable, therefore, that the visual pyramids do not play the part of the simple eyes of the *Vertebrata,* and the only alternative appears to be the adoption of a modification of the theory of *mosaic vision,* propounded many years by Johannes Müller.

Each visual pyramid, isolated from its fellows by its coat of pigment, may be supposed, in fact, to play the part of a very narrow straight tube, with blackened walls, one end of which is turned towards the external world, while the other incloses the extremity of one of the nerve fibres. The only light which can reach the latter, under these circumstances, is such as proceeds from points which lie in the

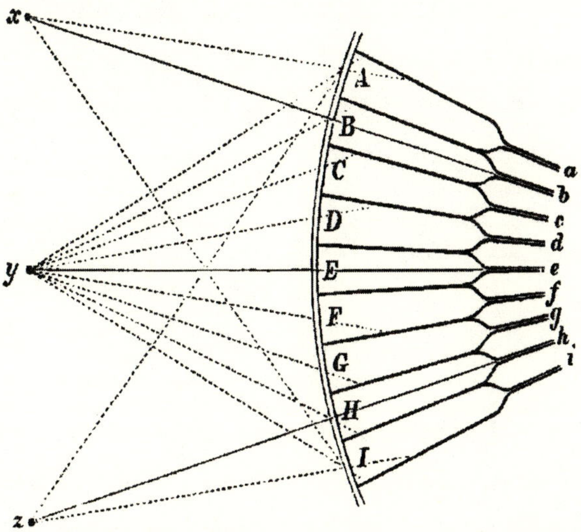

FIG. 29.—Diagram showing the course of rays of light from three points x, y, z, through the nine visual rods (supposed to be empty tubes) A—I of a compound eye; a—i, the nerve fibres connected with the visual rods.

direction of a straight line represented by the produced axis of the tubes.

Suppose A—I to be nine such tubes, a—i the corresponding nerve fibres, and x y z three points from which light proceeds. Then it will be obvious that the only light from *x* which will excite sensation, will be the ray which traverses B and reaches the nerve-fibre *b*, while that from *y* will affect only *e*, and that from *x* only *h*. The result, translated into sensation, will be three points of light on a dark ground, each of which answers to one of the luminous points, and indicates its direction in reference to the eye and its angular distance from the other two.*

The only modification needed in the original form of the theory of mosaic vision, is the supposition that part, or the whole, of the visual rod, is not merely a passive transmitter of light to the nerve-fibre, but is, itself, in someway concerned in transmuting the mode of motion, light, into that other mode of motion which we call nervous energy. The visual rod is, in fact, to be regarded as the physiological end of the nerve, and the instrument by which the conversion of the one form of motion into the other takes place; just as the auditory hairs are instruments by which the sonorous waves are converted into molecular movements of the substance of the auditory nerves.

It is wonderfully interesting to observe that, when the so-called compound eye is interpreted in this manner,

* Since the visual rods are strongly refracting solids, and not empty tubes, the diagram given in fig. 29 does not represent the true course of the rays, indicated by dotted lines, which fall obliquely on any cornea of a crayfish's eye. Such rays will be more or less bent towards the axis of the visual rod of that cornea; but whether they reach its apex and so affect the nerve or not will depend on the curvature of the cornea; its refractive index and that of the crystalline cone; and the relation between the length and the thickness of the latter.

the apparent wide difference between it and the vertebrate eye gives place to a fundamental resemblance. The rods and cones of the retina of the vertebrate eye are extraordinarily similar in their form and their relations to the fibres of the optic nerve, to the visual rods of the arthropod eye. And the morphological discrepancy, which is at first so striking, and which arises from the fact that the free ends of the visual rods are turned towards the light, while those of the rods and cones of the vertebrate eye are turned from it, becomes a confirmation of the parallel between the two when the development of the vertebrate eye is taken into account. For it is demonstrable that the deep surface of the retina in which the rods and cones lie, is really a part of the outer surface of the body turned inwards, in the course of the singular developmental changes which give rise to the brain and the eye of vertebrate animals.

Thus the crayfish has, at any rate, two of the higher sense organs, the ear and the eye, which we possess ourselves; and it may seem a superfluous, not to say a frivolous, question, if any one should ask whether it can hear and see.

But, in truth, the inquiry, if properly limited, is a very pertinent one. That the crayfish is led by the use of its eyes and ears to approach some objects and avoid others, is beyond all doubt; and, in this sense, most indubitably it can both hear and see. But if the question means, do luminous vibrations give it the sensations of light and darkness, of colour and form and distance, which they give to us? and do sonorous vibrations produce the feelings of noise and tone, of melody and of harmony, as in us?—it is by no means to be answered hastily, perhaps cannot be answered at all, except in a tentative, probable way.

The phenomena to which we give the names of sound and colour are not physical things, but are states of consciousness, dependent, there is every reason to believe, on the functional activity of certain parts of our brains. Melody and harmony are names for states of consciousness which arise when at least two sensations of sound have been produced. All these are manufactured articles, products of the human brain; and it would be exceedingly hazardous to affirm that organs capable of giving rise to the same products exist in the vastly simpler nervous system of the crustacean. It would be the height of absurdity to expect from a meat-jack the sort of work which is performed by a Jacquard loom; and it appears to me to be little less preposterous to look for the production of anything analogous to the more subtle phenomena of the human mind in something so minute and rude in comparison to the human brain, as the insignificant cerebral ganglia of the crayfish.

At the most, one may be justified in supposing the existence of something approaching dull feeling in ourselves; and, to return to the problem stated in the beginning of this chapter, so far as such obscure consciousness accompanies the molecular changes of its nervous substance, it will be right to speak of the mind of a crayfish. But it will be obvious that it is merely putting the cart before the horse, to speak of such a mind as a factor in the work done by the organism, when it is merely a dim symbol of a part of such work in the doing.

Whether the crayfish possesses consciousness or not, however, does not affect the question of its being an engine, the actions of which at any moment depend, on the one hand, upon the series of molecular changes excited, either by internal or by external causes, in its neuromuscular machinery; and, on the other, upon the disposition and the properties of the parts of that machinery. And such a self-adjusting machine, containing the immediate conditions of its action within itself, is what is properly understood by an automaton.

Crayfishes, as we have seen, may attain a considerable age; and there is no means of knowing how long they might live, if protected from the innumerable destructive influences to which they are at all ages liable.

It is a widely received notion that the energies of living matter have a natural tendency to decline, and finally disappear; and that the death of the body, as a whole, is the necessary correlate of its life. That all living things sooner or later perish needs no demonstration, but it would be difficult to find satisfactory grounds for the belief that they must needs do so. The analogy of a machine that, sooner or later, must be brought to a standstill by the wear and tear of its parts, does not hold, inasmuch as the animal mechanism is continually renewed and repaired; and, though it is true that individual components of the body are constantly dying, yet their places are taken by vigorous successors. A city remains, notwithstanding the constant death-rate of its inhabitants; and such an organism as a crayfish is only a corporate unity, made up of innumerable partially independent individualities.

Whatever might be the longevity of crayfishes under imaginable perfect conditions, the fact that, notwithstanding the great number of eggs they produce, their number remains pretty much the same in a given district, if we take the average of a period of years, shows that about as many die as are born; and that, without the process of reproduction, the species would soon come to an end.

There are many examples among members of the group of *Crustacea* to which the crayfish belongs, of animals which produce young from internally developed germs, as some plants throw off bulbs which are capable of reproducing the parent stock; such is the case, for example, with the common water flea (*Daphnia*). But nothing of this kind has been observed in the crayfish; in which, as in the higher animals, the reproduction of the species is dependent upon the combination of two kinds of living matter, which are developed in different individuals, termed *males* and *females*.

These two kinds of living matter are *ova* and *spermatozoa*, and they are developed in special organs, the *ovary* and the *testis*. The ovary is lodged in the female; the testis, in the male.

The *ovary* (fig. 30, *ov*) is a body of a trefoil form, which is situated immediately beneath, or in front of, the heart, between the floor of the pericardial sinus and the alimentary canal. From the ventral face of this

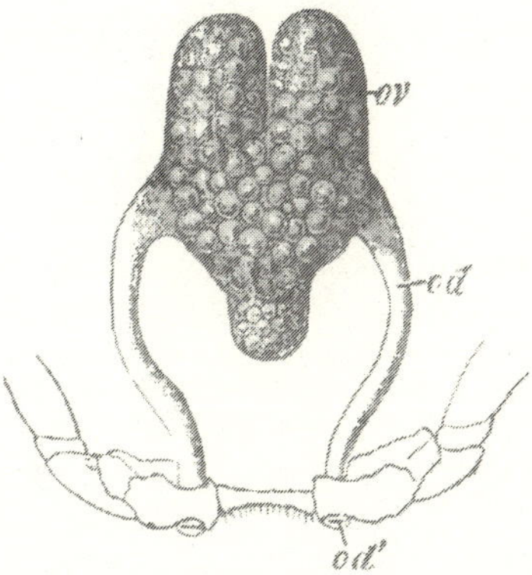

FIG. 30.—*Astacus fluviatilis.*—The female reproductive organs (× 2); *ov*, ovary; *od*, oviduct; *od'*, aperture of oviduct.

organ two short and wide canals, the *oviducts* (*od*), lead down to the bases of the second pair of walking limbs, and terminate in the apertures (*od'*) already noticed there.

The *testis* (fig. 31, *t*) is somewhat similar in form to the ovary, but, the three divisions are much narrower and more elongated: the hinder median division lies under the heart; the anterior divisions are situated between the heart behind, and the stomach and the liver in front (figs. 5 and 12, *t*). From the point at which the

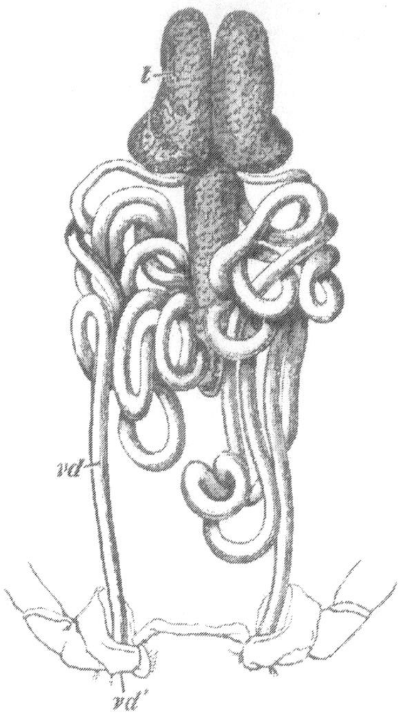

FIG. 31.—*Astacus fluviatilis.*—The male reproductive organs (× 2); *t*, testis; *vd*, vas deferens; *vd'*, aperture of vas deferens.

three divisions join, proceed two ducts, which are termed the *vasa deferentia* (fig. 31, *vd*). These are very narrow, long, and make many coils before they reach the apertures upon the bases of the hindermost pair of walking limbs, by which they open externally (fig. 31, *vd'*, and fig. 35, *vd*). Both the ovary and the testis are very much larger during the breeding season than at other times; the large brownish-yellow eggs become conspicuous in the ovary,

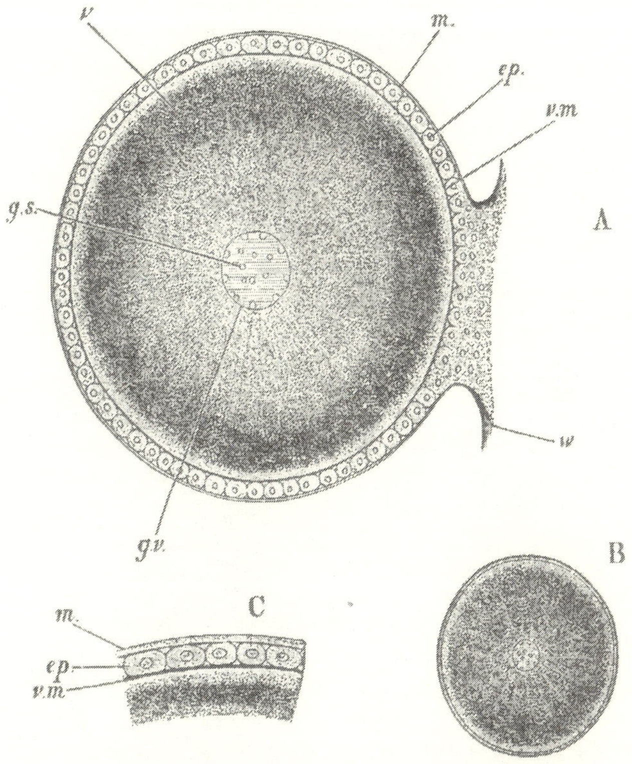

FIG. 32.—*Astacus fluviatilis.*—A, a two-thirds grown egg contained in its ovisac (× 50); B, an egg removed from the ovisac (× 10); C, a portion of the wall of an ovisac with the adjacent portion of the contained egg, highly magnified; *ep*, epithelium of ovisac; *gs*, germinal spots; *gv*, germinal vesicle; *m*, membrana propria; *v*, vitellus; *vm*, vitelline membrane; *w*, stalk of ovisac.

and the testis assumes a milk-white colour, at this period.

The walls of the ovary are lined internally by a layer of

nucleated cells, separated from the cavity of the organ by a delicate structureless membrane. The growth of these cells gives rise to papillary elevations which project into the cavity of the ovary, and eventually become globular

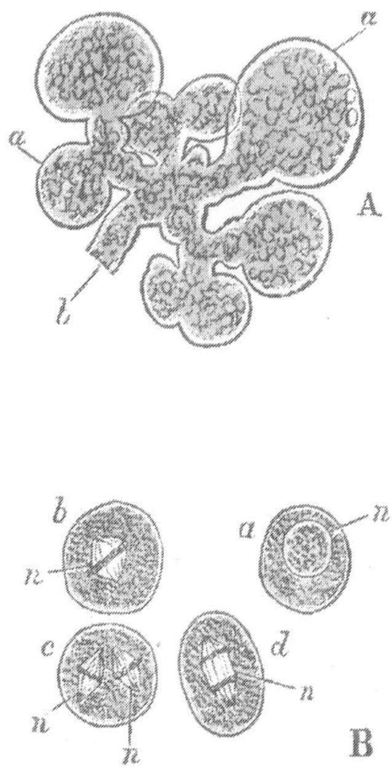

FIG. 33.—*Astacus fluviatilis.*—A, a lobule of the testis, showing *a*, acini, springing from *b*, the ultimate termination of a duct (× 50). B, spermatic cells; *a*, with an ordinary globular nucleus *n*; *b*, with a spindle-shaped nucleus; *c*, with two similar nuclei; and *d*, with a nucleus undergoing division (× 600).

bodies attached by short stalks, and invested by the structureless membrane as a *membrana propria* (fig. 32, *m*). These are the *ovisacs*. In the mass of cells which becomes the ovisac, one rapidly increases in size and occupies the centre of the ovisac, while the others surround it as a peripheral coat (*ep.*). This central cell is the *ovum.* Its nucleus enlarges, and becomes what is called the *germinal vesicle* (*g.v.*). At the same time numerous small corpuscles, flattened externally and convex internally, appear in it and are the *germinal spots* (*g.s.*). The protoplasm of the cell, as it enlarges, becomes granular and opaque, assumes a deep brownish-yellow colour, and is thus converted into the *yelk* or *vitellus* (*v.*). As the egg grows, a structureless *vitelline membrane* is formed between the vitellus and the cells which line the ovisac, and incloses the egg, as in a bag. Finally, the ovisac bursts, and the egg, falling into the cavity of the ovary, makes its way down the oviduct, and sooner or later passes out by its aperture. When they leave the oviduct, the ova are invested by a viscous, transparent substance, which attaches them to the swimmerets of the female, and then sets; thus each egg, inclosed in a tough case, is firmly suspended by a stalk, which, on the one side, is continued into the substance of the case, while, on the other, it is fixed to the swimmeret. The swimmerets are kept constantly in motion, so that the eggs are well supplied with aërated water.

The testis consists of an immense number of minute spheroidal vesicles (fig. 33, A, *a*), attached like grapes to the ends of short stalks (*b*), formed by the ultimate ramifications of the vasa deferentia. The vesicles may, in fact, be regarded as dilatations of the ends and sides

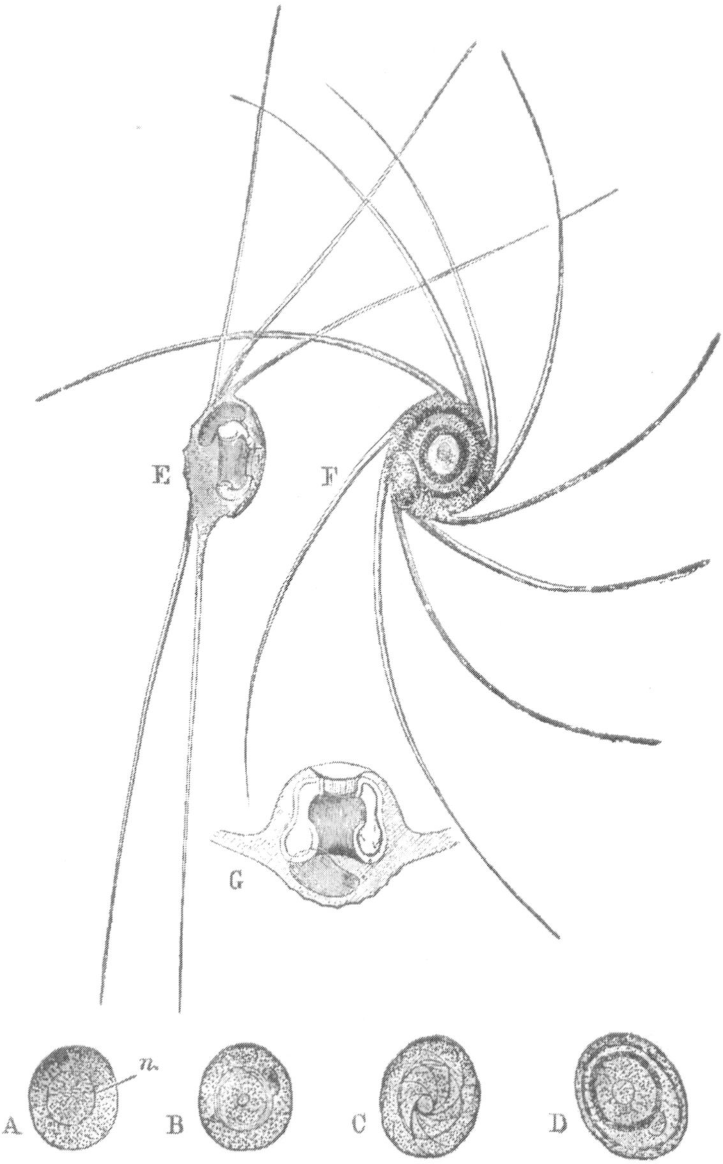

FIG. 34.—*Astacus fluviatilis.*—A—D, different stages in the development of a spermatozoon from a seminal cell; E, a mature spermatozoon seen from the side; F, the same viewed *en face* (all × 850); G, a diagrammatic vertical section of the same.

of the finest branches of the ducts of the testis. The cavity of each vesicle is filled by the large nucleated cells which line its walls (fig. 33, B), and, as the breeding season approaches, these cells multiply by division. Finally, they undergo some very singular changes of form and internal structure (fig. 34, A—D), each becoming converted into a flattened spheroidal body, about $\frac{1}{1700}$th of an inch in diameter, provided with a number of slender curved rays, which stand out from its sides (fig. 34, E—G). These are the *spermatozoa*.

The spermatozoa accumulate in the testicular vesicles, and give rise to a milky-looking substance, which traverses the smaller ducts, and eventually fills the vasa deferentia. This substance, however, consists, in addition to the spermatozoa, of a viscid material, secreted by the walls of the vasa deferentia, which envelopes the spermatozoa, and gives the secretion of the testis the form and the consistency of threads of vermicelli.

The ripening and detachment of both the ova and the spermatozoa take place immediately after the completion of ecdysis in the early autumn; and at this time, which is the breeding season, the males seek the females with great avidity, in order to deposit the fertilizing matter contained in the vasa deferentia on the sterna of their hinder thoracic and anterior abdominal somites. There it adheres as a whitish, chalky-looking mass; but the manner in which the contained spermatozoa reach and enter the ova is unknown. The analogy of what occurs in other animals, however, leaves no doubt that an actual mixture of the male and female elements takes place and constitutes the essential part of the process of impregnation.

Ova to which spermatozoa have had no access, give rise to no progeny; but, in the impregnated ovum, the young crayfish takes its origin in a manner to be described below, when the question of development is dealt with.

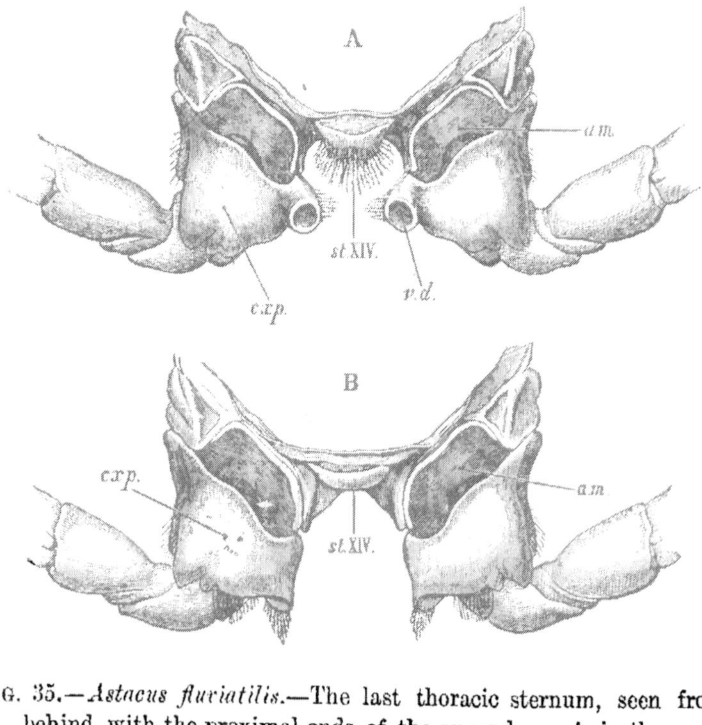

FIG. 35.—*Astacus fluviatilis.*—The last thoracic sternum, seen from behind, with the proximal ends of the appendages, A, in the male, B, in the female, (× 3). *am*, articular membrane; *cxp*, coxopodite; *st XIV*, last thoracic sternum; *vd*, aperture of vas deferens.

CHAPTER IV.

THE MORPHOLOGY OF THE COMMON CRAYFISH: THE STRUCTURE AND THE DEVELOPMENT OF THE INDIVIDUAL.

In the two preceding chapters the crayfish has been studied from the point of view of the physiologist, who, regarding an animal as a mechanism, endeavours to discover how it does that which it does. And, practically, this way of looking at the matter is the same as that of the teleologist. For, if all that we know concerning the purpose of a mechanism is derived from observation of the manner in which it acts, it is all one, whether we say that the properties and the connexions of its parts account for its actions, or that its structure is adapted to the performance of those actions.

Hence it necessarily follows that physiological phenomena can be expressed in the language of teleology. On the assumption that the preservation of the individual, and the continuance of the species, are the final causes of the organization of an animal, the existence of that organization is, in a certain sense, explained, when it is shown that it is fitted for the attainment of those ends; although, perhaps, the importance of demonstrating the proposition that a thing is fitted to do that which it does, is not very great.

But whatever may be the value of teleological explanations, there is a large series of facts, which have as yet been passed over, or touched only incidentally, of which they take no account. These constitute the subject matter of *Morphology*, which is related to physiology much as, in the not-living world, crystallography is related to the study of the chemical and physical properties of minerals.

Carbonate of lime, for example, is a definite compound of calcium, carbon, and oxygen, and it has a great variety of physical and chemical properties. But it may be studied under another aspect, as a substance capable of assuming crystalline forms, which, though extraordinarily various, may all be reduced to certain geometrical types. It is the business of the crystallographer to work out the relations of these forms; and, in so doing, he takes no note of the other properties of carbonate of lime.

In like manner, the morphologist directs his attention to the relations of form between different parts of the same animal, and between different animals; and these relations would be unchanged if animals were mere dead matter, devoid of all physiological properties—a kind of mineral capable of a peculiar mode of growth.

A familiar exemplification of the difference between teleology and morphology may be found in such works of human art as houses.

A house is certainly, to a great extent, an illustration of adaptation to purpose, and its structure is, to that extent, explicable by teleological reasonings. The roof and the walls are intended to keep out the weather; the foundation is meant to afford support and to exclude damp; one room is contrived for the purpose of a kitchen; another for that of a coal-cellar; a third for that of a dining-room; others are constructed to serve as sleeping rooms, and so on; doors, chimneys, windows, drains, are all more or less elaborate contrivances directed towards one end, the comfort and health of the dwellers in the house. What is sometimes called sanitary architecture, now-a-days, is based upon considerations of house teleology. But though all houses are, to begin with and essentially, means adapted to the ends of shelter and comfort, they may be, and too often are, dealt with from a point of view, in which adaptation to purpose is largely disregarded, and the chief attention of the architect is given to the form of the house. A house may be built in the Gothic, the Italian, or the Queen Anne style; and a house in any one of these styles of architecture may be just as convenient or inconvenient, just as well or as ill adapted to the wants of the resident therein, as any of the others. Yet the three are exceedingly different.

To apply all this to the crayfish. It is, in a sense, a house with a great variety of rooms and offices, in which the work of the indwelling life in feeding, breathing, moving, and reproducing itself, is done. But the same may be said of the crayfish's neighbours, the perch and the water-snail; and they do all these things neither better nor worse, in relation to the conditions of their existence, than the crayfish does. Yet the most cursory inspection is sufficient to show that the "styles of architecture" of the three are even more widely different than are those of the Gothic, Italian, and Queen Anne houses.

That which Architecture, as an art conversant with pure form, is to buildings, Morphology, as a science conversant with pure form, is to animals and plants. And we may now proceed to occupy ourselves exclusively with the morphological aspect of the crayfish.

As I have already mentioned, when dealing with the physiology of the crayfish, the entire body of the animal, when reduced to its simplest morphological expression, may be represented as a cylinder, closed at each end, except so far as it is perforated by the alimentary apertures (fig. 6); or we may say that it is a tube, inclosing another tube, the edges of the two being continuous at their extremities. The outer tube has a chitinous outer coat or cuticle, which is continued on to the inner face of the inner tube. Neglecting this for the present, the outermost part of the wall of the outer tube, which answers to the *epidermis* of the higher animals, and the innermost part of the wall of the inner tube, which is an *epithelium*, are formed by a layer of nucleated cells. A continuous layer of cells, therefore, is everywhere to be found on both the external and the internal free surfaces of the body. So far as these cells belong to the proper external wall of the body, they constitute the *ectoderm*, and so far as they belong to its proper internal wall, they compose the *endoderm*. Between these two layers of nucleated cells lie all the other parts of the body, composed of connective tissue, muscles, vessels, and nerves; and all these (with the exception of the ganglionic chain, which we shall see properly belongs to the ectoderm) may be regarded as a single thick stratum, which, as it lies between the ectoderm and the endoderm, is called the *mesoderm*.

If the intestine were closed posteriorly instead of opening by the vent, the crayfish would virtually be an elongated sac, with one opening, the mouth, affording an entrance into the alimentary cavity: and, round this cavity, the three layers just referred to—endoderm, mesoderm, and ectoderm—would be disposed concentrically.

We have seen that the body of the crayfish thus composed is obviously separable into three regions—the *cephalon* or head, the *thorax*, and the *abdomen*. The latter is at once distinguished by the size and the mobility of its segments: while the thoracic region is marked off from that of the head, outwardly, only by the cervical groove. But, when the carapace is removed, the lateral depression already mentioned, in which the scaphognathite lies, clearly indicates the natural boundary between the head and the thorax. It has further been observed that there are, in all, twenty pairs of appendages, the six hindermost of which are attached to the abdomen. If the other fourteen pairs are carefully removed, it will be found that the six anterior belong to the head, and the eight posterior to the thorax.

The abdominal region may now be studied in further detail. Each of its seven movable segments, except the telson, represents a sort of morphological unit, the repetition of which makes up the whole fabric of the body.

If the abdomen is divided transversely between the

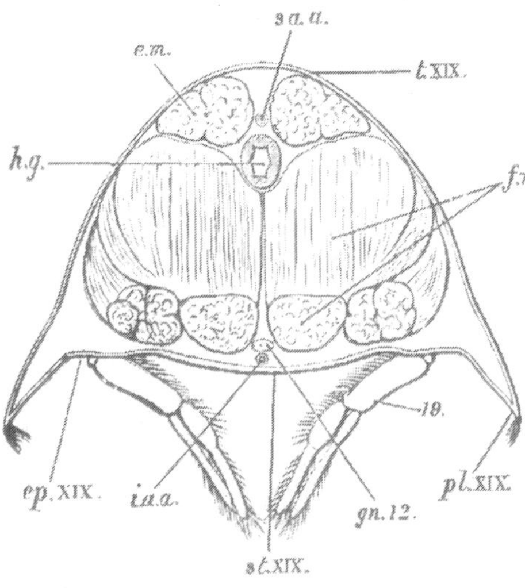

FIG. 36.—*Astacus fluviatilis.*—A transverse section through the nineteenth (fifth abdominal) somite (× 2). *e.m.*, extensor muscles; *f.m.*, flexor muscles; *gn. 12*, the fifth abdominal ganglion; *h.g.*, hind-gut; *i.a.a.*, inferior abdominal artery; *s.a.a*, superior abdominal artery; *pl. XIX*, pleura of the somite; *st. XIX*, its sternum; *t. XIX*, its tergum; *ep. XIX*, its epimera; *19*, its appendages.

fourth and fifth, and the fifth and sixth segments, the fifth will be isolated, and can be studied apart. It constitutes what is called a *metamere;* in which are distinguishable a central part termed the *somite,* and two *appendages* (fig. 36).

In the exoskeleton of the somites of the abdomen several regions have already been distinguished; and although they constitute one continuous whole, it will be convenient to speak of the *sternum* (fig. 36, *st. XIX*), the *tergum* (*t. XIX*), and, the *pleura* (*pl. XIX*), as if they were separate parts, and to distinguish that portion of the sternal region, which lies between the articulation of the appendage and the pleuron, on each side, as the *epimeron* (*ep. XIX*). Adopting this nomenclature, it may be said of the fifth somite of the abdomen, that it consists of a segment of the exoskeleton, divisible into tergum, pleura, epimera, and sternum, with which two appendages are articulated; that it contains a double ganglion (*gn. 12*), a section of the flexor (*fm*) and extensor (*em*) muscles, and of the alimentary (*hg*) and vascular (*s.a.a, i.a.a*) systems.

The appendage (fig. 36, *19*), which is attached to an articular cavity situated between the sternum and the epimeron, is seen to consist of a stalk or stem, which is made up of a very short basal joint, the *coxopodite* (fig. 37, D and E, *cx.p*), followed by a long cylindrical second joint, the *basipodite* (*b.p*), and receives the name of *protopodite.* At its free end, it bears two flattened narrow

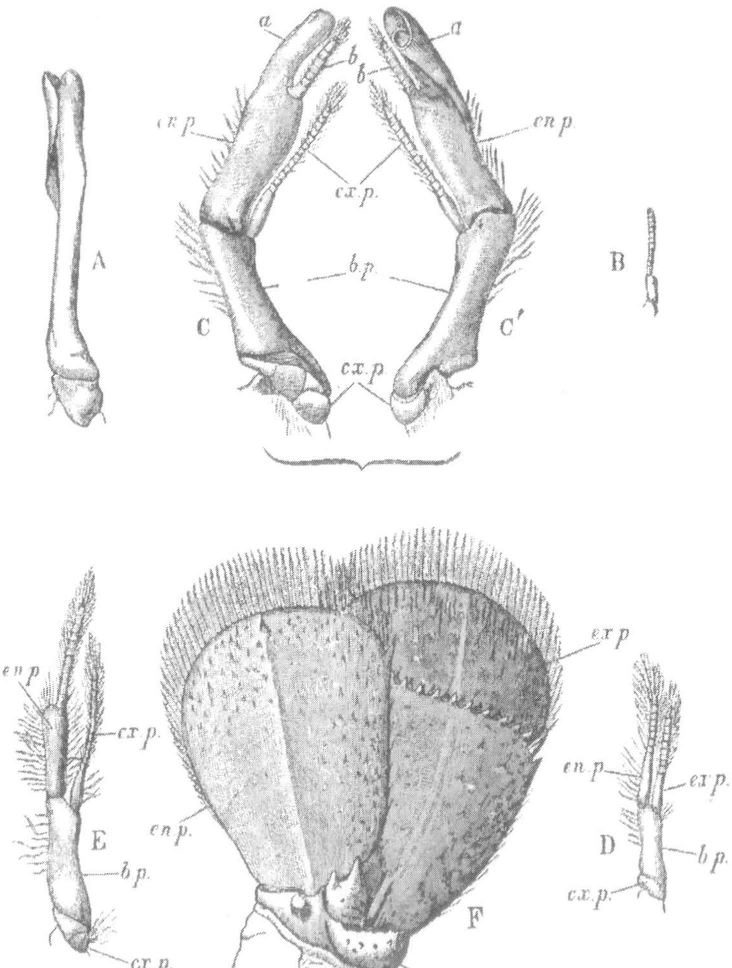

FIG. 37.—*Astacus fluviatilis.*—Appendages of the left side of the abdomen (× 3). A, the posterior face of the first appendage of the male; B, the same of the female; C, posterior, and C′, anterior faces of the second appendage of the male; D, the third appendage of the male; E, the same of the female; F, the sixth appendage. *a*, the rolled plate of the endopodite; *b*, the jointed extremity of the same; *bp.*, basipodite; *cx.p.*, coxopodite; *en.p.*, endopodite; *ex.p.*, exopodite.

plates, of which one is attached to the inner side of the extremity of the protopodite, and is called the *endopodite* (*en.p*), while the other is fixed a little higher up to the outer side of that extremity, and is the *exopodite* (*ex.p*). The exopodite is shorter than the endopodite. The endopodite is broad and is undivided for about half its length, from the attached end; the other half is narrower, and is divided into a number of small segments, which, however, are not united by definite articulations, but are merely marked off from one another by slight constrictions of the exoskeleton. The exopodite has a similar structure, but its undivided portion is shorter and narrower. The edges of both the exopodite and the endopodite are fringed with long setæ.

In the female crayfish, the appendages of this and of the fourth and third somites are larger than in the male (compare *D* and *E*, fig. 37).

The fourth and fifth somites, with their appendages, may be described in the same terms as the third, and in the sixth there is no difficulty in recognising the corresponding parts of the somite; but the appendages (fig. 37, *F*), which constitute the lateral portions of the caudal fin, at first sight appear very different. In their size, no less than in their appearance, they depart widely from the appendages of the preceding somites. Nevertheless, each will be found to consist of a basal stalk, answering to the protopodite (*cx.p*), which however is very broad and thick, and is not divided into two joints; and of two terminal oval plates, which represent the endopodite (*en.p*) and the exopodite (*ex.p*). The latter is divided by a transverse suture into two pieces; and the edge of the larger or basal moiety is beset with short spines, of which two, at the outer end of the series, are larger than the rest.

The second somite is longer than the first (fig. 1); it has very broad pleura, while those of the first somite are small and hidden by the overlapping front margins of the pleura of the second somite.

In the female, the appendages of the second somite of the abdomen are similar to those of the third, fourth, and fifth somites; but in those of the first somite (fig. 37, *B*), there is a considerable variation. Sometimes, in fact, the appendages of this somite are altogether wanting; sometimes one is present, and not the other; and sometimes both are found. But, when they exist, these appendages are always small; and the protopodite is followed by only one imperfectly jointed filament, which appears to represent the endopodite of the other appendages.

In the male, the appendages of the first and second somites of the abdomen are not only of relatively large size, but they are widely different from the rest, those of the first somite departing from the general type further than those of the second. In the latter (*C*, *C′*) there is a protopodite (*cx.p*, *bp*) with the ordinary structure, and it is followed by an endopodite (*en.p*) and an exopodite (*ex.p*); but the former is singularly modified. The undivided basal part is large, and is produced on the inner side into a lamella (*a*), which extends slightly beyond the end of the terminal jointed portion (*b*). The inner half of this lamella is rolled upon itself, in such a manner as to give rise to a hollow cone, something like an extinguisher (*C′*, *a*).

The appendage of the first somite (*A*) is an unjointed styliform body, which appears to represent the protopodite, together with the basal part and the inner prolongation of the endopodite of the preceding appendage. The terminal half of the appendage is really a broad plate, slightly bifid at the summit, but the sides of the plate are rolled in, in such a manner that the anterior half bends round and partially incloses the posterior half. They thus give rise to a canal, which is open at each end, and only partially closed behind.

These two pairs of curiously modified appendages are ordinarily turned forwards and applied against the sterna of the posterior part of the thorax, in the interval between the bases of the hinder thoracic limbs (see fig. 3, *A*). They serve as conduits by which the spermatic matter of the male is conveyed from the openings of the ducts of the testes to its destination.

If we confine our attention to the third, fourth, and fifth metameres of the abdomen of the crayfish, it is obvious that the several somites and their appendages, and the various regions or parts into which they are divisible, correspond with one another, not only in form, but in their relations to the general plan of the whole abdomen. Or, in other words, a diagrammatic plan of one somite will serve for all the three somites, with insignificant variations in detail. The assertion that these somites are constructed upon the same plan, involves no more hypothesis than the statement of an architect, that three houses are built upon the same plan, though the façades and the internal decorations may differ more or less.

In the language of morphology, such conformity in the plan of organisation is termed *homology*. Hence, the several metameres in question and their appendages, are *homologous* with one another; while the regions of the somites, and the parts of their appendages, are also *homologues*.

When the comparison is extended to the sixth metamere, the homology of the different parts with those of the other metameres, is undeniable, notwithstanding the great differences which they present. To recur to a previous comparison, the ground plan of the building is the same, though the proportions are varied. So with regard to the first and second metameres. In the second pair of appendages of the male, the difference from the ordinary type of appendage is comparable to that produced by adding a portico or a turret to the building; while, in the first pair of appendages of the female, it is as if one wing of the edifice were left unbuilt;

and, in those of the male, as if all the rooms were run into one.

It is further to be remarked, that, just as of a row of houses built upon the same plan, one may be arranged so as to serve as a dwelling-house, another as a warehouse, and another as a lecture hall, so the homologous appendages of the crayfish are made to subserve various functions. And as the fitness of the dwelling-house, the warehouse, and the lecture-hall for their several purposes would not in the least help us to understand why they should all be built upon the same general plan; so, the adaptation of the appendages of the abdomen of the crayfish to the discharge of their several functions does not explain why those parts are homologous. On the contrary, it would seem simpler that each part should have been constructed in such a manner as to perform its allotted function in the best possible manner, without reference to the rest. The proceedings of an architect, who insisted on constructing every building in a town on the plan of a Gothic cathedral, would not be explicable by considerations of fitness or convenience.

In the cephalothorax, the division into somites is not at first obvious, for, as we have seen, the dorsal or tergal surface is covered over by a continuous shield, distinguished into thoracic and cephalic regions only by the cervical groove. Even here, however, when a transverse section of the thorax is compared with that of the abdomen (figs. 15 and 36), it will be obvious that the tergal and the sternal regions of the two answer to one another; while the branchiostegites correspond with greatly developed pleura; and the inner wall of the branchial chamber, which extends from the bases of the appendages to the attachment of the branchiostegite, represents an immensely enlarged epimeral region.

On examination of the sternal aspect of the cephalothorax the signs of division into somites become plain (figs. 3 and 39, *A*). Between the last two ambulatory limbs there is an easily recognisable sternum (*XIV.*), though it is considerably narrower than any of the sterna of the abdominal somites, and differs from them in shape.

The deep transverse fold which separates this hindermost thoracic sternum from the rest of the sternal wall of the cephalothorax, is continued upwards on the inner or epimeral wall of the branchial cavity; and thus the sternal and the epimeral portions of the posterior thoracic somite are naturally marked off from those of the more anterior somites.

The epimeral region of this somite presents a very curious structure (fig. 38). Immediately above the articular cavities for the appendages there is a shield-shaped plate, the posterior, convex edge of which is sharp, prominent, and setose. Close to its upper boundary the plate exhibits a round perforation (*plb.*), to the margins of which the stem of the hindermost pleurobranchia (fig. 4, *plb. 14*) is attached; and in front of this, it is connected, by a narrow neck, with an elongated triangular piece, which takes a vertical direction, and lies in the fold which separates the posterior thoracic somite from the next in front. The base of this

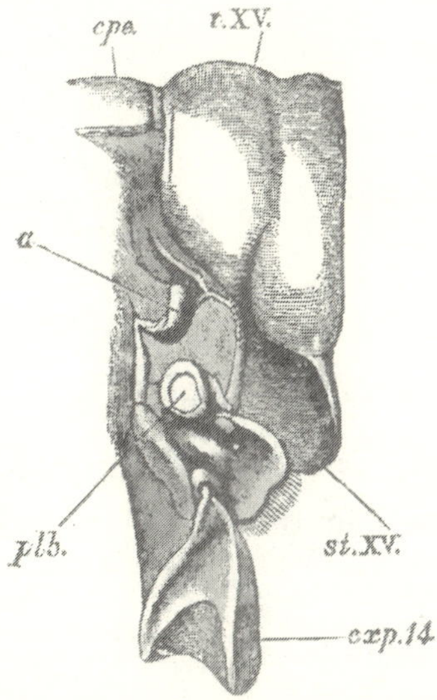

FIG. 38.—*Astacus fluviatilis.*—The mode of connexion between the last thoracic and the first abdominal somites (× 3). *a*, L-shaped bar; *cpe*, carapace; *cxp. 14*, coxopodite of the last ambulatory leg; *plb.*, place of attachment of the pleurobranchia; *st. XV*, sternum, and *t. XV*, tergum of the first abdominal somite.

piece unites with the epimeron of the penultimate somite. Its apex is connected with the anterior end of the horizontal arm of an L-shaped calcified bar (fig. 38, *a*), the upper end of the vertical arm of which is firmly, but moveably, connected with the anterior and lateral edge of the tergum of the first abdominal somite (*t. XV.*). The tendon of one of the large extensor muscles of the abdomen is attached close to it.

The sternum and the shield-shaped epimeral plates constitute a solid, continuously calcified, ventral element of the skeleton, to which the posterior pair of legs is attached; and as this structure is united with the somites in front of and behind it only by soft cuticle, except where the shield-shaped plate is connected, by the intermediation of the triangular piece, with the epimeron which lies in front of it, it is freely movable backwards and forwards on the imperfect hinge thus constituted.

In the same way, the first somite of the abdomen, and, consequently, the abdomen as a whole, moves upon the hinges formed by the union of the L-shaped pieces with the triangular pieces.

In the rest of the thorax, the sternal and the epimeral regions of the several somites are all firmly united together. Nevertheless, shallow grooves answering to folds of the cuticle, which run from the intervals between the articular cavities for the limbs towards the tergal end of the inner wall of the branchial chamber, mark off the epimeral portions of as many somites as there are sterna, from one another.

A short distance above the articular cavities a transverse groove separates a nearly square area of the lower part of the epimeron from the rest. Towards the anterior and upper angle of this area, in the two somites which lie immediately in front of the hindermost, there is a small round aperture for the attachment of the

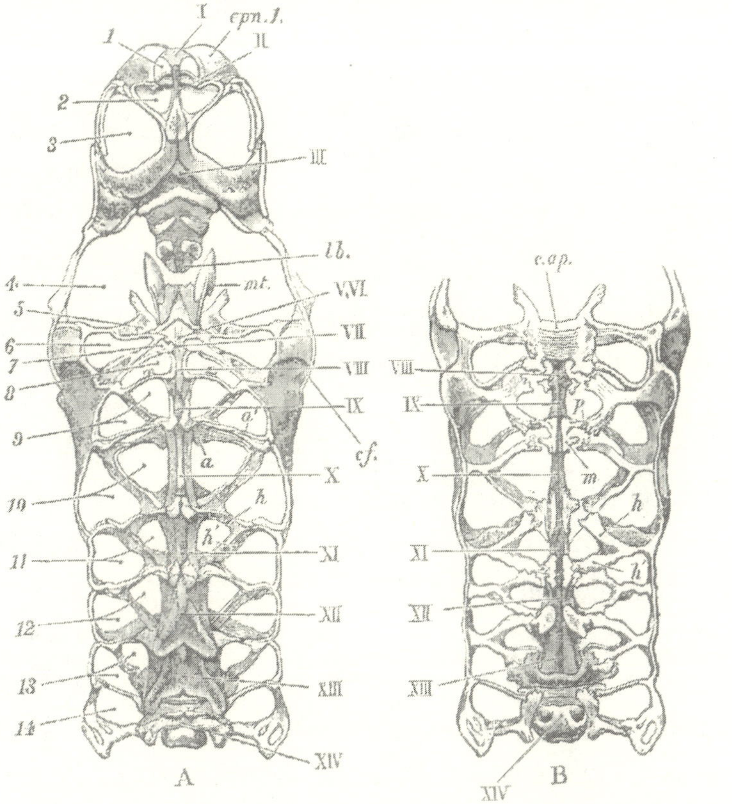

FIG. 39.—*Astacus fluviatilis.*—The cephalothoracic sterna and the endophragmal system (× 2). *A*, from beneath; *B*, from above. *a*, *a'*, arthrophragms or partitions between the articular cavities for the limbs; *c.ap*, cephalic apodeme; *cf*, cervical fold; *epn. 1*, epimeron of the antennulary somite; *h*, anterior, and *h'*, posterior horizontal process of endopleurite; *lb*, labrum; *m*, mesophragm; *mt*, metastoma; *p*, paraphragm; *I—XIV*, cephalothoracic sterna; *1—14*, articular cavities of the cephalothoracic appendages. (The anterior cephalic sterna are bent downwards in A so as to bring them into the same plane with the remaining cephalothoracic sterna; in B these sterna are not shown.)

rudimentary branchia. These areæ of the epimera, in fact, correspond with the shield-shaped plate of the hindermost somite. In the next most anterior somite (that which bears the first pair of ambulatory legs) there is only a small elevation in the place of the rudimentary branchia; and in the anterior four thoracic somites nothing of the kind is visible.

On the sternal aspect of the thorax (figs. 3 and 39, A) a triangular space is interposed between the basal joints or coxopodites of the penultimate and the ante-penultimate pairs of ambulatory legs, while the coxopodites of the more anterior limbs are closely approximated. The triangular area in question is occupied by two sterna (fig. 39, A, *XII*, *XIII*), the lateral margins of which are raised into flange-like ridges. The next two sterna (*X*, *XI*) are longer, especially that which lies between the forceps (*X*), but they are very narrow; while the lateral processes are reduced to mere tubercles at the posterior ends of the sterna. Between the three pairs of maxillipedes, the sterna (*VII*, *VIII*, *IX*) are yet narrower, and become gradually shorter; but traces of the tubercles at their posterior ends are still discernible. The most anterior of these sternal rods passes into a transversely elongated plate, shaped like a broad arrow (*V*, *VI*), which is constituted by the conjoined sterna of the two posterior somites of the head.

Anteriorly to this, and between it and the posterior end of the elongated oral aperture, the sternal region is occupied only by soft or imperfectly calcified cuticle, which, on each side of the hinder part of the mouth, passes into one of the lobes of the metastoma (*mt*). At the base of each of these lobes there is a calcified plate, united by an oblique suture with another, which occupies the whole length of the lobe and gives it firmness. The soft narrow lip which constitutes the lateral boundary of the oral aperture, and lies between it and the mandible, passes, in front, into the posterior face of the labrum (*lb*).

In front of the mouth, the sternal region which appertains, in part, to the antennæ, and, in part, to the mandibles, is obvious as a broad plate (*III*), termed the *epistoma*. The middle third of the posterior edge of the epistoma gives rise to a thickened transverse ridge, with rounded ends, slightly excavated behind, and is then continued into the labrum (*lb*), which is strengthened by three pairs of calcifications, arranged in a longitudinal series. The sides of the front edge of the epistoma are excavated, and bound the articular cavities for the basal joints of the antennæ (*3*); but, in the middle line, the epistoma is continued forwards into a spear-head shaped process (figs. 39 and 40, *II*), to which the posterior end of the antennulary sternum contributes. The antennulary sternum is very narrow, and its anterior or upper end runs into a small but distinct conical median spine (fig. 40, *t.*). Upon this follows an uncalcified plate, bent into the form of a half cylinder (*I*), which lies between the inner ends of the eye-stalks and is united with adjacent parts only by flexible cuticle, so that it is freely movable. This represents the whole of the sternal region, and probably more, of the ophthalmic somite.

The sterna of fourteen somites are thus identifiable in the cephalothorax. The corresponding epimera are

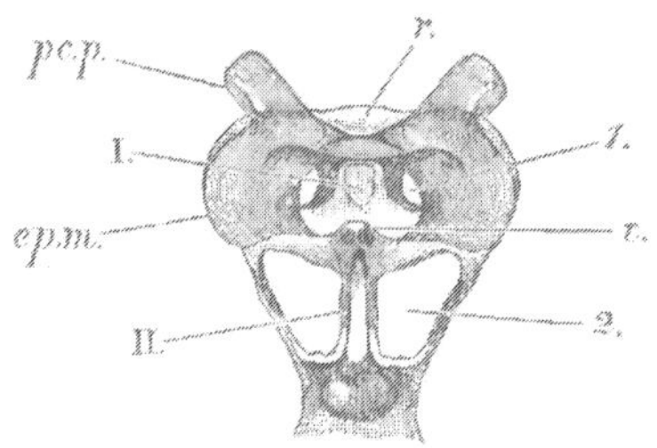

FIG. 40.—*Astacus fluviatilis.*—The ophthalmic and antennulary somites (× 3). *I*, ophthalmic, and *II*, antennulary sternum; *1*, articular surface for eyestalk; *2*, for antennule; *epm*, epimeral plate; *pcp*, procephalic process; *r*, base of rostrum; *t*, tubercle.

represented, in the thorax, by the thin inner walls of the branchial chamber; the pleura, by the branchiostegites; and the terga, by so much of the median region of the carapace as lies behind the cervical groove. That part of the carapace which is situated in front of this groove occupies the place of the terga of the head; while the low ridge, skirting the oral and præ-oral region, in which it terminates laterally, represents the pleura of the cephalic somites.

The epimera of the head are, for the most part, very narrow; but those of the antennulary somite are broad plates (fig. 40, *epm.*), which constitute the posterior wall of the orbits. I am inclined to think that a transverse ridge, which unites these under the base of the rostrum, represents the tergum of the antennulary somite, and that the rostrum itself belongs to the next or antennary somite.*

The sharp convex ventral edge of the rostrum (fig. 41) is produced into a single, or sometimes two divergent spines, which descend, in front of the ophthalmic somite, towards the conical tubercle mentioned above: it thus gives rise to an imperfect partition between the orbits.

FIG. 41.—*Astacus fluviatilis.*—The rostrum, seen from the left side.

The internal face of the sternal wall of the whole of the thorax and of the post-oral part of the head, presents a complicated arrangement of hard parts, which is known as the *endophragmal system* (figs. 39, *B*, 42, and 43), and which performs the office of an internal skeleton by affording attachment to muscles, and serving to protect important viscera, while at the same time it ties the somites together, and unites them into a solid whole. In reality, however, the curious pillars and bulkheads which enter into the composition of the endophragmal system are all

* There are some singular marine crustacea, the *Squillidæ*, in which both the ophthalmic and the antennary somites are free and movable, while the rostrum is articulated with the tergum of the antennary somite.

mere infoldings of the cuticle, or *apodemes;* and, as such, they are shed along with the other cuticular structures during the process of ecdysis.

Without entering into unnecessary details, the general principle of the construction of the endophragmal skeleton may be stated as follows. Four apodemes are developed between every two somites, and as every apodeme is a fold of the cuticle, it follows that the anterior wall of each belongs to the somite in front, and the posterior wall to the somite behind. All four apodemes lie in the ventral half of the somite and form a single transverse series; consequently there are two nearer the middle line, which are termed the *endosternites,* and two further off, which are the *endopleurites.* The former lie at the inner, and the latter at the outer ends of the partitions or *arthrophragms* (fig. 39, A, *a*, *a′*, fig. 42, *aph*), between the articular cavities for the basal joints of the limbs, and they spring partly from the latter and partly from the sternum and the epimera respectively.

The endosternite (fig. 42, *ens.*) ascends vertically, with a slight inclination forwards, and its summit narrows and assumes the form of a pillar, with a flat, transversely elongated capital. The inner prolongation of the capital is called the *mesophragm* (*mph.*), the outer the *paraphragm* (*pph.*). The mesophragms of the two endosternites of a somite usually unite by a median suture, and thus form a complete arch over the sternal canal (*s.c.*), which lies between the endosternites.

The endopleurites (*en.pl.*) are also vertical plates, but they are relatively shorter, and their inner angles give off two nearly horizontal processes, one of which passes obliquely forwards (fig. 39, B, *h*, fig. 42, *h.p.*) and unites with the paraphragm of the endosternite of the somite in front, while the other, passing obliquely backwards (fig. 39, *h′*), becomes similarly connected with the endosternite of the somite behind.

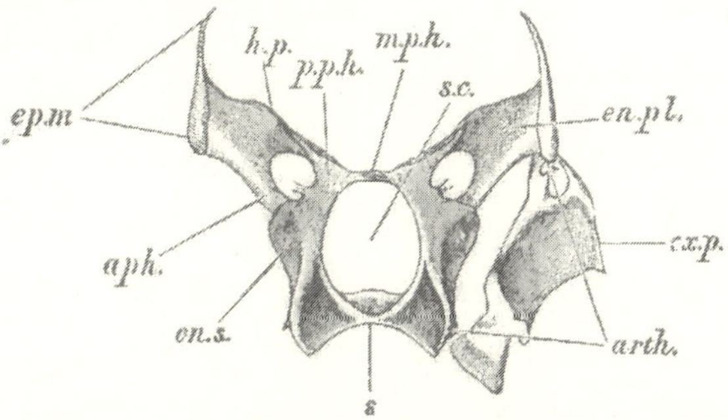

FIG. 42.—*Astacus fluviatilis.*—A segment of the endophragmal system (× 3). *aph*, arthrophragm; *arth*, arthrodial or articular cavity; *cxp*, coxopodite of the ambulatory leg; *enpl*, endopleurite; *ens*, endosternite; *epm*, epimeron; *hp*, horizontal process of endopleurite; *mph*, mesophragm; *pph*, paraphragm; *s*, sternum of somite; *sc*, sternal canal.

The endopleurites of the last thoracic somite are rudimentary, and its endosternites are small. On the other hand, the mesophragmal processes of the endosternites of the two posterior somites of the head (fig. 39, B, *c.ap*), by which the endophragmal system terminates in front, are particularly strong and closely united together. They thus, with their endopleurites, form a solid partition between the stomach, which lies upon them, and the mass of coalesced anterior thoracic and posterior cephalic ganglia situated beneath them. Strong processes are given off from their anterior and outer angles, which curve round the tendons of the adductor muscles of the mandibles, and give attachment to the abductors.

In front of the mouth there is no such endophragmal system as that which lies behind it. But the anterior gastric muscles are attached to two flat calcified plates, which appear to lie in the interior of the head (though they are really situated in its upper and front wall) on each side of the base of the rostrum, and are called the *procephalic processes* (figs. 40, 43, *p.cp*). Each of these plates constitutes the posterior wall of a narrow cavity which opens externally into the roof of the orbit, and has been regarded (though, as it appears to me, without sufficient reason) as an olfactory organ. I am disposed to think, though I have not been able to obtain complete evidence of the fact, that the procephalic processes are the representatives of the "procephalic lobes" which terminate the anterior end of the body in the embryo crayfish. At any rate, they occupy the same position relatively to the eyes and to the carapace; and the hidden position of these processes, in the adult, appears to arise from the extension of the carapace at the base of the rostrum over the fore part of the originally free sternal surface of the head. It has thus covered over the procephalic processes, in which the sternal wall of the body terminated; and the cavities which lie in front of them are simply the interspaces left between the inferior or posterior wall of the prolongation of the carapace and the originally exposed external faces of these regions of the cephalic integument.

Fourteen somites having thus been distinguished in the cephalothorax, and six being obvious in the abdomen, it is clear that there is a somite for every pair of appendages. And, if we suppose the carapace divided into segments answering to these sterna, the whole body will be made up of twenty somites, each having a pair of appendages. As the carapace, however, is not actually divided into terga in correspondence with the sterna which it covers, all we can safely conclude from the anatomical facts is that it represents the tergal region of the somites, not that it is formed by the coalescence of primarily distinct terga. In the head, and in the greater part of the thorax, the somites are, as it were, run together, but the last thoracic somite is partly free and to a slight extent moveable, while the abdominal somites are all free, and moveably articulated together. At the anterior end of the body, and, apparently, from the antennary somite, the tergal region gives rise to the rostrum, which projects between and beyond the eyes. At the opposite extremity, the telson is a corresponding median outgrowth of the last somite, which has become moveably articulated therewith. The narrowing of the sternal moieties of the anterior thoracic somites, together with the sudden widening of the same parts in the posterior cephalic somites, gives rise to the lateral depression (fig. 39, *cf*) in which the scaphognathite lies. The limit thus indicated corresponds with that marked by the cervical groove upon the surface of the carapace, and separates the head from the thorax. The three pair of maxillipedes (*7, 8, 9*), the forceps (*10*), the ambulatory

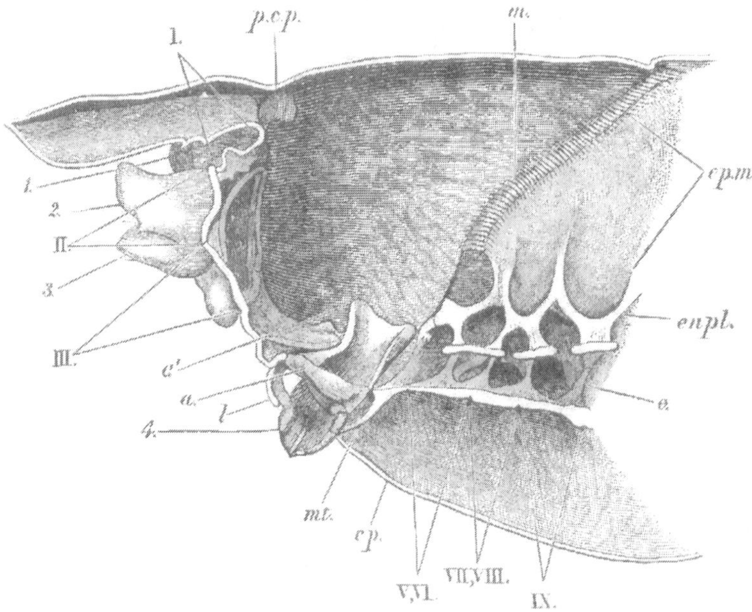

FIG. 43.—*Astacus fluviatilis.*—Longitudinal section of the anterior part of the cephalothorax (× 3). *I—IX*, sterna of first nine cephalothoracic somites; *1*, eyestalk; *2*, basal joint of antennule; *3*, basal joint of antenna; *4*, mandible; *a*, inner division of the masticatory surface of the mandible; *a'*, apophysis of the mandible for muscular attachment; *cp*, free edge of carapace; *e*, endosternite; *enpl*, endopleurite; *cpm*, epimeral plate; *l*, labrum; *m*, muscular fibres connecting epimera with interior of carapace; *mt*, metastoma; *pcp*. procephalic process.

limbs (*11*–*14*), and the eight somites of which they are the appendages (*VII*—*XIV*), lie behind this boundary and belong to the thorax. The two pairs of maxillæ (*5*, *6*) the mandibles (*4*), the antennæ (*3*), the antennules (*2*), the eyestalks (*1*), and the six somites to which they are attached (*I*—*VI*), lie in front of the boundary and compose the head.

Another important point to be noticed is that, in front of the mouth, the sternum of the antennary somite (fig. 43, *III*) is inclined at an angle of 60° or 70° to the direction of the sterna behind the mouth. The sternum of the antennulary somite (*II*) is at right angles to the latter; and that of the eyes (*I*) looks upwards as well as forwards. Hence, the front of the head beneath the rostrum, though it looks forwards, or even upwards, is homologous with the sternal aspect of the other somites. It is for this reason that the feelers and the eyestalks take a direction so different from that of the other appendages. The change of aspect of the sternal surface in front of the mouth, thus effected, is what is termed the *cephalic flexure.*

Since the skeleton which invests the trunk of the crayfish is made up of a twenty-fold repetition of somites, homologous with those of the abdomen, we may expect to find that the appendages of the thorax and of the head, however unlike they may seem to be to those of the abdomen, are nevertheless reducible to the same fundamental plan.

The third maxillipede is one of the most complete of these appendages, and may be advantageously made the starting point of the study of the whole series.

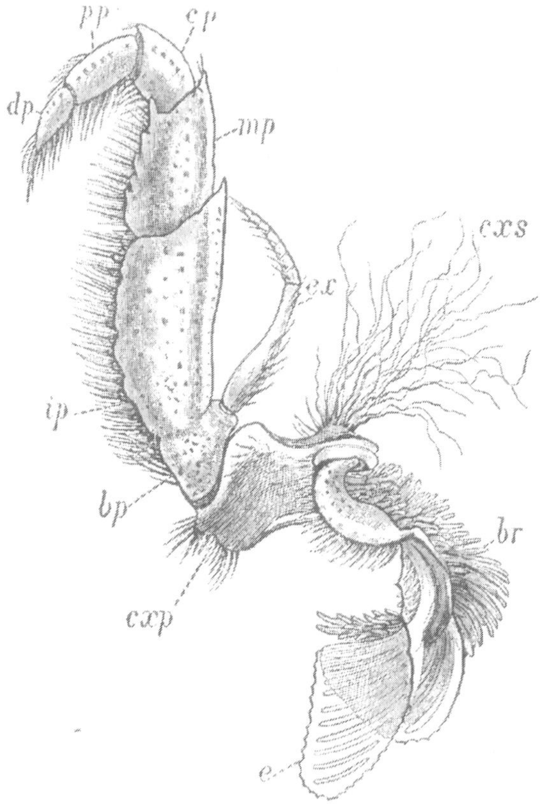

FIG. 44.—*Astacus fluviatilis.*—The third or external maxillipede of the left side (× 3). *e*, lamina, and *br*, branchial filaments of the podobranchia; *cxp*, coxopodite; *cxs*, coxopoditic setæ; *bp*, basipodite; *ex*, exopodite; *ip*, ischiopodite; *mp*, meropodite; *cp*, carpopodite; *pp*, propodite; *dp*, dactylopodite.

Neglecting details for the moment, it may be said that the appendage consists of a basal portion (fig. 44, *cxp*, *bp*), with two terminal divisions (*ip* to *dp*, and *ex*), which are directed forwards, below the mouth, and a third, lateral appendage (*e*, *br*), which runs up, beneath the carapace, into the branchial chamber. The latter is the gill, or podobranchia, attached to this limb, and it is something not represented in the abdominal limbs. But, with regard to the rest of the maxillipede, it is obvious that the basal portion (*cxp*, *bp*) represents the protopodite, and the two terminal divisions the endopodite and the exopodite respectively. It has been observed that, in the abdominal appendages, the extent to which segmentation occurs in homologous parts varies indefinitely; an endopodite, for example, may be a continuous plate, or may be subdivided into many joints. In the maxillipede, the basal portion is divided into two joints; and, as in the abdominal limb, the first, or that which articulates with the thorax, is termed the *coxopodite* (*cxp*), while the second is the *basipodite* (*bp*). The stout, leg-like endopodite appears to be the direct continuation of the basipodite; while the much more narrow and slender exopodite articulates with its outer side. The exopodite (*ex*) is by no means unlike one of the exopodites of the abdominal limbs, consisting as it does of an undivided base and a many-jointed terminal filament. The endopodite, on the contrary, is strong and massive, and is divided into five joints, named, from that nearest to the base onwards, *ischiopodite* (*ip*), *meropodite* (*mp*), *carpopodite* (*cp*), *propodite* (*pp*), and *dactylopodite* (*dp*).

The second maxillipede (fig. 45, B) has essentially the same composition as the first, but the exopodite (*ex*) is relatively larger, the endopodite (*ip*—*dp*) smaller and softer; and, while the ischiopodite (*ip*) is the longest joint in the third maxillipede, it is the meropodite (*mp*) which is longest in the second. In the first maxillipede

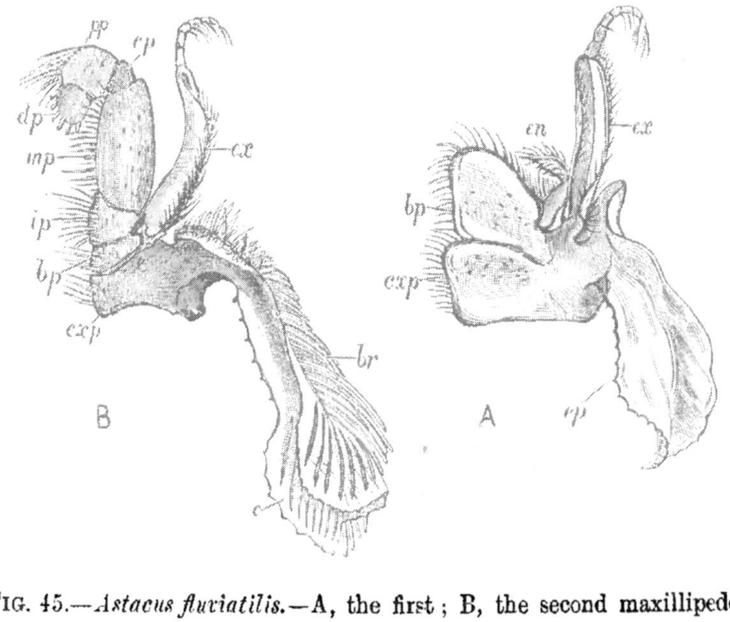

FIG. 45.—*Astacus fluviatilis.*—A, the first; B, the second maxillipede of the left side (× 3). *cxp*, coxopodite; *bp*, basipodite; *c*, *br*, podobranchia; *ep*, epipodite; *en*, endopodite; *ex*, exopodite; *ip*, ischiopodite; *mp*, meropodite; *cp*, carpopodite; *pp*, propodite; *dp*, dactylopodite.

(fig. 45, A) a great modification has taken place. The coxopodite (*cxp*) and the basipodite (*bp*) are broad thin plates with setose cutting edges, while the endopodite (*en*) is short and only two-jointed, and the undivided portion of the exopodite (*ex*) is very long. The place of the podobranchia is taken by a broad soft membranous plate entirely devoid of branchial filaments (*ep*). Thus, in the series of the thoracic limbs, on passing forwards from the third maxillipede, we find that though the plan of the appendages remains the same; (1) the protopodite increases in relative size; (2) the endopodite diminishes; (3) the exopodite increases; (4) the podobranchia finally takes the form of a broad membranous plate and loses its branchial filaments.

Writers on descriptive Zoology usually refer to the parts of the maxillipedes under different names from those which are employed here. The protopodite and the endopodite taken together are commonly called the *stem* of the maxillipede, while the exopodite is the *palp*, and the metamorphosed podobranchia, the real nature of which is not recognised, is termed the *flagellum*.

When the comparison of the maxillipedes with the abdominal members, however, had shown the fundamental uniformity of composition of the two, it became desirable to invent a nomenclature of the homologous parts which should be capable of a general application. The names of protopodite, endopodite, exopodite, which I have adopted as the equivalents of the "stem" and the "palp," were proposed by Milne-Edwards, who at the same time suggested *epipodite* for the "flagellum." And the lamellar process of the first maxillipede is now very generally termed an epipodite; while the podobranchiæ, which have exactly the same relations to the following limbs, are spoken of as if they were totally different structures, under the name of branchiæ or gills.

The flagellum or epipodite of the first maxillipede, however, is nothing but the slightly modified stem of a podobranchia, which has lost its branchial filaments; but the term "epipodite" may be conveniently used for podobranchiæ thus modified. Unfortunately, the same term is applied to certain lamelliform portions of the branchiæ of other crustacea, which answer to the laminæ of the crayfishes' branchiæ; and this ambiguity must be borne in mind, though it is of no great moment.

On examining an appendage from that part of the thorax which lies behind the third maxillipede, say, for example, the sixth thoracic limb (the second walking leg) (fig. 46), the two joints of the protopodite and the five joints of the endopodite are at once identifiable, and so is the podobranchia; but the exopodite has vanished altogether. In the eighth, or last, thoracic limb, the podobranchia has also disappeared. The fifth and sixth limbs also differ from the seventh and eighth, in being chelate; that is to say, one angle of the distal end of the propodite is prolonged and forms the fixed leg of the pincer. The produced angle is that which is turned downwards when the limb is fully extended (fig. 46). In the forceps, the great chela is formed in just the same way; the only important difference lies in the fact that, as in the external maxillipede, the basipodite and the ischiopodite are immoveably united. Thus,

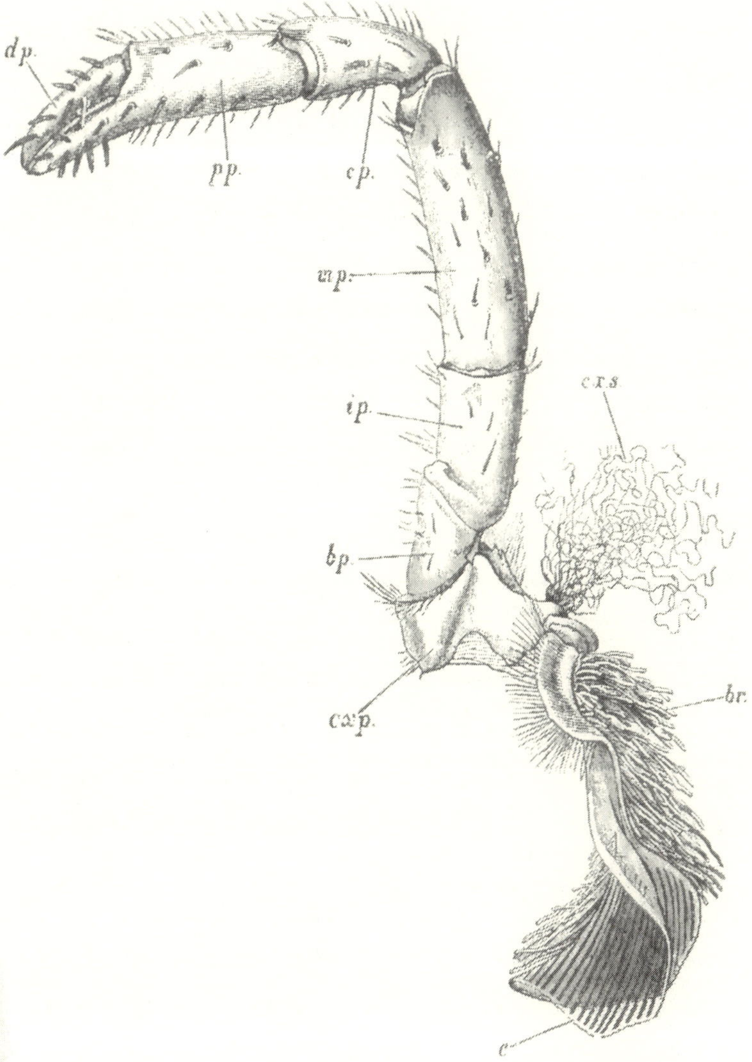

FIG. 46.—*Astacus fluviatilis.*—The second ambulatory leg of the left side (× 3). *cxp*, coxopodite ; *bp*, basipodite ; *br*, gill ; *cxs*, coxopoditic setæ ; *e*, lamina of gill or epipodite ; *ip*, ischiopodite ; *mp*, meropodite ; *cp*, carpopodite ; *pp*, propodite ; *dp*, dactylopodite.

the limbs of the thorax are all reducible to the same type as those of the abdomen, if we suppose that, in the posterior five pair, the exopodites are suppressed; and that, in all but the last, podobranchiæ are superadded.

Turning to the appendages of the head, the second maxilla (fig. 47, C) presents a further modification of the disposition of the parts seen in the first maxillipede. The coxopodite (*cxp*) and the basipodite (*bp*) are still thinner and more lamellar, and are subdivided by deep fissures which extend from their inner edges. The endopodite (*en*) is very small and undivided. In the place of the exopodite and the epipodite there is only one great plate, the scaphognathite (*sg*) which either is such an epipodite as that of the first maxillipede with its anterior basal process much enlarged, or represents both the exopodite and the epipodite. In the first maxilla (B), the exopodite and the epipodite have disappeared, and the endopodite (*en*) is insignificant and unjointed. In the mandibles (A), the representative of the protopodite is strong and transversely elongated. Its broad inner or oral end presents a semicircular masticatory surface divided by a deep longitudinal groove into two toothed ridges. The one of these follows the convex anterior or inferior contour of the masticatory surface, projects far beyond the other, and is provided with a sharp serrated edge; the other (fig. 43, *a*) gives rise to the straight posterior or superior contour of the masticatory surface, and is more obtusely tuberculated. In front, the inner ridge is continued into a process by which the mandible articulates with the epistoma (fig. 47, A, *ar*). The endo-

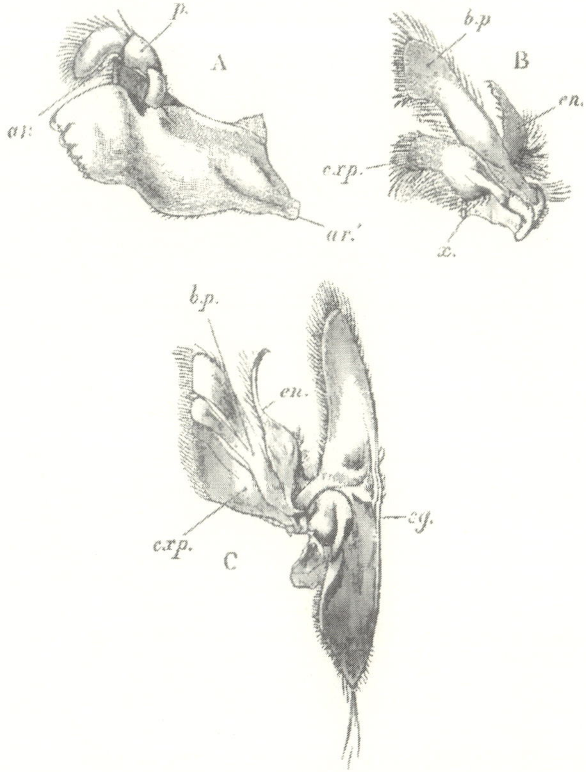

FIG. 47.—*Astacus fluviatilis.*—A, mandible; B, first maxilla; C, second maxilla of the left side (× 3). *ar*, internal, and *ar'*, external articular process of the mandible; *bp*, basipodite; *cxp*, coxopodite; *en*, endopodite; *p*, palp of the mandible; *sg*, scaphognathite; *x*, internal process of the first maxilla.

podite is represented by the three-jointed *palp* (*p*), the terminal joint of which is oval and beset with numerous strong setæ, which are especially abundant along its anterior edge.

In the antenna (fig. 48, C) the protopodite is two-jointed. The basal segment is small, and its ventral face presents the conical prominence on the posterior aspect of which is the aperture of the duct of the renal gland (*gg*). The terminal segment is larger and is subdivided by deep longitudinal folds, one upon the dorsal and

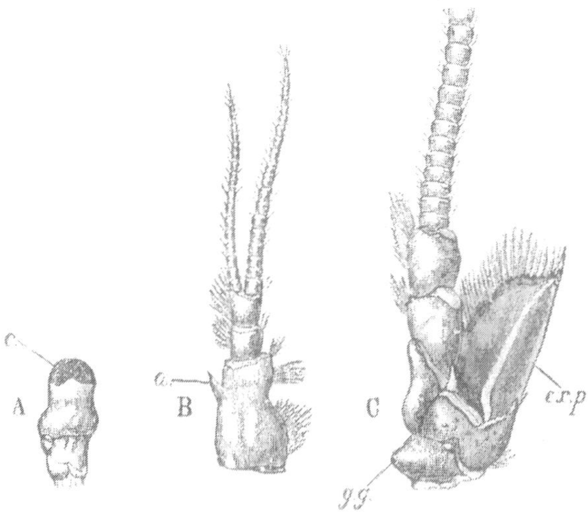

FIG. 48.—*Astacus fluviatilis.*—A, eye-stalk; B, antennule; C, antenna of the left side (× 3). *a*, spine of the basal joint of the antennule; *c*, corneal surface of the eye: *exp*, exopodite or squame of the antenna; *gg*, aperture of the duct of the green gland.

one upon the ventral face, into two moieties which are more or less moveable upon one another. In front and externally it bears the broad flat *squame* (*exp*) of the antenna, as an exopodite. Internally, the long annulated "feeler" which represents the endopodite, is connected with it by two stout basal segments.

The antennule (fig. 48, B) has a three-jointed stem and two terminal annulated filaments, the outer of which is thicker and longer than the inner, and lies rather above as well as external to the latter. The peculiar form of the basal segment of the stem of the antennule has already been adverted to (p. 116). It is longer than the other two segments put together, and near the anterior end its sternal edge is produced into a single strong spine (*a*). The stem of the antennule answers to the protopodite of the other limbs, though its division into three joints is unusual; the two terminal annulated filaments represent the endopodite and the exopodite.

Finally, the eyestalk (A) has just the same structure as the protopodite of an abdominal limb, having a short basal and a long cylindrical terminal joint.

From this brief statement of the characters of the appendages, it is clear that, in whatever sense it is allowable to say that the appendages of the abdomen are constructed upon one plan, which is modified in execution by the excess of development of one part over another, or by the suppression of parts, or by the coalescence of one part with another, it is allowable to say that all the appendages are constructed on the same plan, and are modified on similar principles. Given a general type of appendage consisting of a protopodite, bearing a podobranchia, an endopodite and an exopodite, all the actual appendages are readily derivable from that type.

In addition, therefore, to their adaptation to the purposes which they subserve, the parts of the skeleton of the crayfish show a unity in diversity, such as, if the animal were a piece of human workmanship, would lead us to suppose that the artificer was under an obligation not merely to make a machine capable of doing certain kinds of work, but to subordinate the nature and arrangement of the mechanism to certain fixed architectural conditions.

The lesson thus taught by the skeletal organs is reiterated and enforced by the study of the nervous and the muscular systems. As the skeleton of the whole body is capable of resolution into the skeletons of twenty separate metameres, variously modified and combined; so is the entire ganglionic chain resolvable into twenty pairs of ganglia various in size, distant in this region and approximated in that; and so is the muscular system of the trunk conceivable as the sum of twenty *myotomes* or segments of the muscular system appropriate to a metamere, variously modified according to the degree of mobility of the different regions of the organism.

The building up of the body by the repetition and the modification of a few similar parts, which is so obvious from the study of the general form of the somites and of their appendages, is still more remarkably illustrated, if we pursue our investigations further, and trace out the more intimate structure of these parts. The tough, outer coat, which has been termed the *cuticula*, except so far as it presents different degrees of hardness, from the presence or absence of calcareous salts, is obviously everywhere of the same nature; and, by macerating a crayfish in caustic alkali, which destroys all its other components of the body, it will be readily enough seen that a continuation of the cuticular layer passes in at the mouth and the vent, and lines the alimentary canal; furthermore, that processes of the cuticle covering various parts of the trunk and limbs extend inwards, and afford surfaces of attachment to the muscles, as the *apodemata* and *tendons*. In technical language, the cuticular substance which thus enters so largely into the composition of the bodily fabric of the crayfish is called a *tissue*.

The flesh, or *muscle*, is another kind of tissue, which is readily enough distinguished from cuticular tissue by the naked eye; but, for a complete discrimination of all the different tissues, recourse must be had to the microscope, the application of which to the study of the ultimate optical characters of the morphological constituents of the body has given rise to that branch of morphology which is known as *Histology*.

If we count every formed element of the body, which is separable from the rest by definite characters, as a tissue, there are no more than eight kinds of such tissues in the crayfish; that is to say, every solid constituent of the body consists of one or more of the following eight histological groups:—

1. Blood corpuscles; 2. Epithelium; 3. Connective tissue; 4. Muscle; 5. Nerve; 6. Ova; 7. Spermatozoa; 8. Cuticle.

1. A drop of freshly-drawn blood of the crayfish contains multitudes of small particles, the *blood corpuscles*,

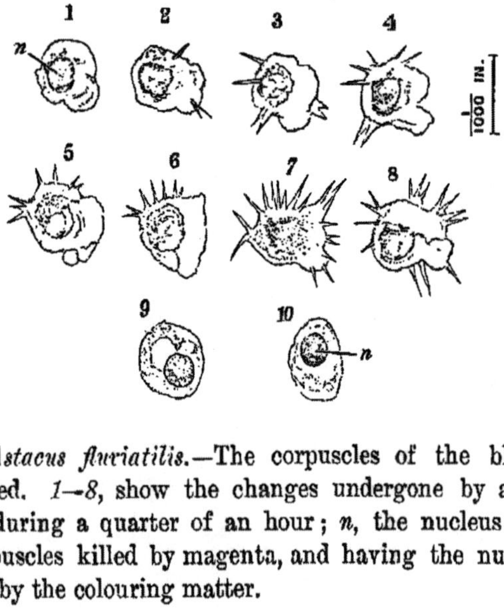

FIG. 49.—*Astacus fluviatilis.*—The corpuscles of the blood, highly magnified. *1–8*, show the changes undergone by a single corpuscle during a quarter of an hour; *n*, the nucleus; *9* and *10* are corpuscles killed by magenta, and having the nucleus deeply stained by the colouring matter.

which rarely exceed 1-700th, and usually are about 1-1000th, of an inch in diameter (fig. 49). They are sometimes pale and delicate, but generally more or less dark, from containing a number of minute strongly refracting granules, and they are ordinarily exceedingly irregular in form. If one of them is watched continuously for two or three minutes, its shape will be seen to undergo the constant but slow changes to which passing reference has already been made (p. 69). One or other of the irregular prolongations will be drawn in, and another thrown out elsewhere. The corpuscle, in fact, has an inherent contractility, like one of those low organisms, known as an *Amœba*, whence its motions are frequently called *amœbiform*. In its interior, an ill-marked oval contour may be seen, indicating the presence of a spheroidal body, about 1-2000th of an inch in diameter, which is the nucleus of the corpuscle (*n*). The addition of some re-agents, such as dilute acetic acid, causes the corpuscles at once to assume a spherical shape, and renders the nucleus very conspicuous (fig. 49, *9* and *10*). The blood corpuscle is, in fact, a simple nucleated cell, composed of a contractile protoplasmic mass, investing a nucleus; it is suspended freely in the blood; and, though as much a part of the crayfish organism as any other of its histological elements, leads a quasi-independent existence in that fluid.

2. Under the general name of *epithelium*, may be included a form of tissue, which everywhere underlies the exoskeleton (where it corresponds with the epidermis of the higher animals), and the cuticular lining of the alimentary canal, extending thence into the hepatic cæca. It is further met with in the generative organs, and in the green gland. Where it forms the subcuticular layer of the integument and of the alimentary canal, it is found to

consist of a protoplasmic substance (fig. 50), in which close set nuclei (*n*) are imbedded. If a number of blood corpuscles could be supposed to be closely aggregated together into a continuous sheet, they would give rise to such a structure as this; and there can be no doubt that it really is an aggregate of nucleated cells, though the limits between the individual cells are rarely visible in the fresh state. In the liver, however, the cells grow, and become detached from one another in the wider and lower

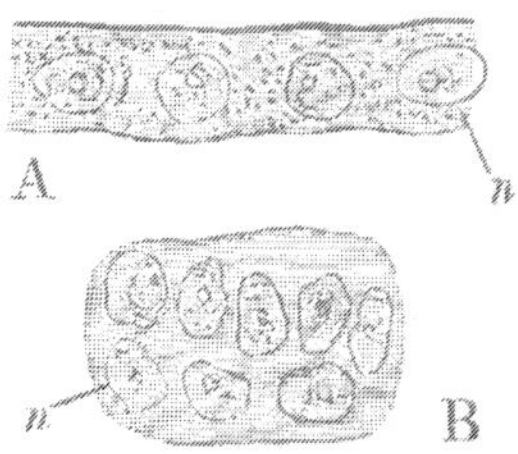

FIG. 50.—*Astacus fluviatilis.*—Epithelium, from the epidermic layer subjacent to the cuticle, highly magnified. *A*, in vertical section; *B*, from the surface. *n*, nuclei.

parts of the cæca, and their essential nature is thus obvious.

3. Immediately beneath the epithelial layer follows a tissue, disposed in bands or sheets, which extend to the subjacent parts, invest them, and connect one with another. Hence this is called *connective tissue.*

The connective tissue presents itself under three forms. In the first there is a transparent homogeneous-looking matrix, or ground substance, through which are scattered many nuclei. In fact, this form of connective tissue very closely resembles the epithelial tissue, except that the intervals between the nuclei are wider, and that the substance in which they are imbedded cannot be broken up into a separate cell-body for each nucleus. In the second form (fig. 51, *A*) the matrix exhibits fine wavy parallel lines, as if it were marked out into imperfect

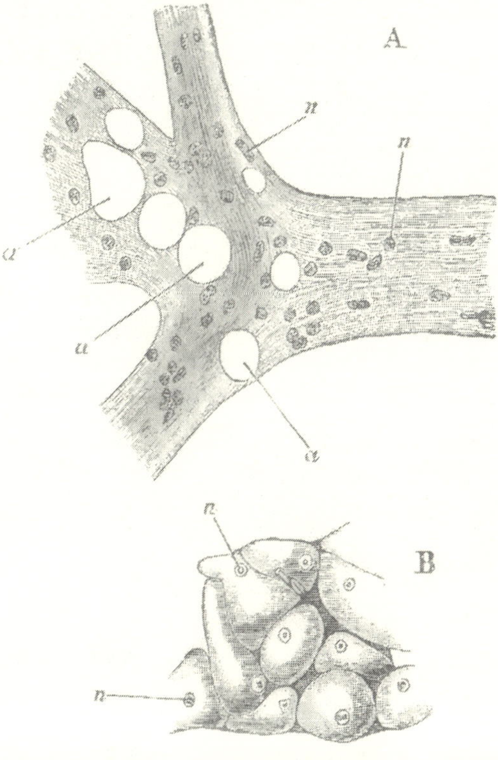

FIG. 51.—*Astacus fluviatilis.*—Connective tissue ; *A*, second form ; *B*, third form. *a*, cavities ; *n*, nuclei. Highly magnified.

fibres. In this form, as in the next to be described, more or less spherical cavities, which contain a clear fluid, are excavated in the matrix; and the number of these is sometimes so great, that the matrix is proportionally very much reduced, and the structure acquires a close superficial similarity to that of the parenchyma of plants. This is still more the case with a third form, in which the matrix itself is marked off into elongated or rounded masses, each of which has a nucleus in its interior (fig. 51, *B*). Under one form or another, the connective tissue extends throughout the body, ensheathing the various organs, and forming the walls of the blood sinuses.

The third form is particularly abundant in the outer investment of the heart, the arteries, the alimentary canal, and the nervous centres. About the cerebral and anterior thoracic ganglia, and on the exterior of the heart, it usually contains more or less fatty matter. In these regions, many of the nuclei, in fact, are hidden by the accumulation round them of granules of various sizes, some of which are composed of fat, while others consist of a proteinaceous material. These aggregates of granules are usually spheroidal; and, with the matrix in which they are imbedded and the nucleus which they surround, they are often readily detached when a portion of the connective tissue is teased out, and are then known as *fat cells*. From what has been said respecting the distribution of the connective tissue, it is obvious that if all the other tissues could be removed, this tissue would form a continuous whole, and represent a sort of model, or cast, of the whole body of the crayfish.

4. The *muscular tissue* of the crayfish always has the form of bands or fibres, of very various thickness, marked, when viewed by transmitted light, by alternate darker and

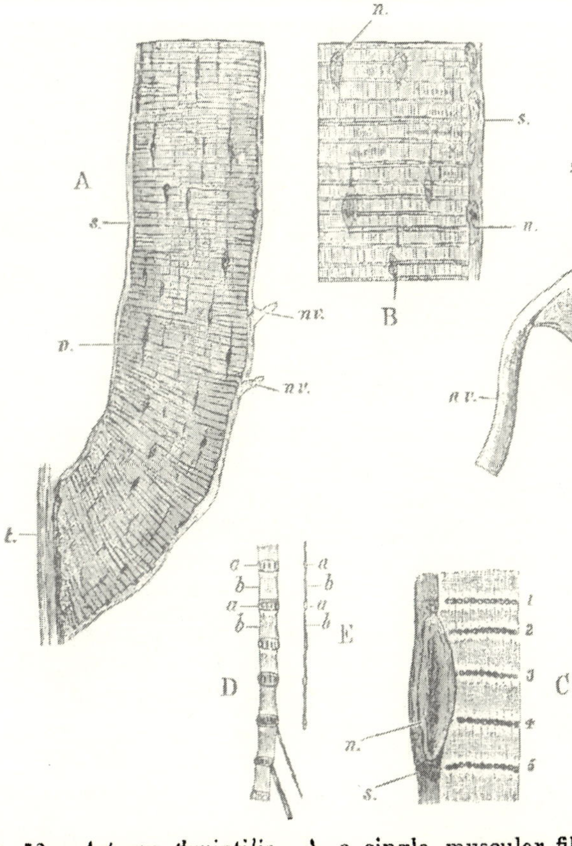

FIG. 52.—*Astacus fluviatilis*.—A, a single muscular fibre, transverse diameter $\frac{1}{110}$th of an inch; B, a portion of the same more highly magnified; C, a smaller portion treated with alcohol and acetic acid still more highly magnified; D and E, the splitting up of a part of a fibre, treated with picro-carmine, into fibrillæ; F, the connection of a nervous with a muscular fibre which has been treated with alcohol and acetic acid. *a*, darker, and *b*, clearer portions of the fibrillæ; *n*, nuclei; *nv*, nerve fibre; *s*, sarcolemma; *t*, tendon; 1—5, successive dark granular striæ answering to the granular portions, *a*, of each fibrilla.

lighter striæ, transversely to the axis of the fibres (fig. 52 A). The distance of the transverse striæ from one another varies with the condition of the muscle, from 1-4,000th of an inch in the quiescent state to as little as 1-30,000th of an inch in that of extreme contraction. The more delicate muscular fibres, like those of the heart and those of the intestine, are imbedded in the connective tissue of the organ, but have no special sheaths.

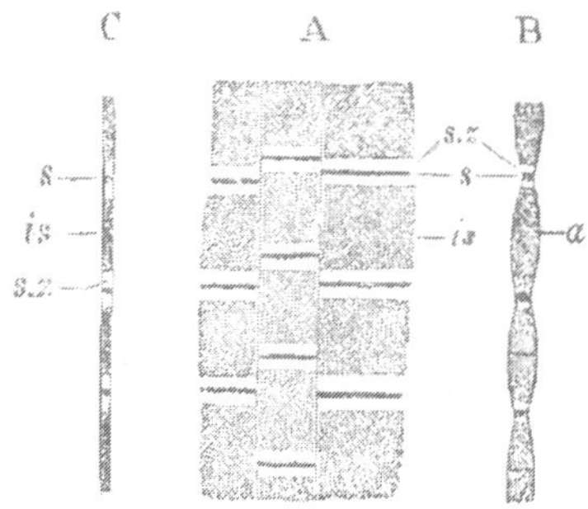

FIG. 53.—*Astacus fluviatilis.*—*A*, living muscular fibres very highly magnified; *B*, a fibrilla treated with solution of sodium chloride; *C*, a fibrilla treated with strong nitric acid. *s*, septal lines; *sz*, septal zones; *is*, interseptal zones; *a*, transverse line in the interseptal zone.

The fibres which make up the more conspicuous muscles of the trunk and limbs, on the other hand, are much larger, and are invested by a thin, transparent, structureless sheath, which is termed the *sarcolemma*. Nuclei are scattered, at intervals, through the striated substance of the muscle; and, in the larger muscular fibres, a layer of nucleated protoplasm lies between the sarcolemma and the striated muscle substance.

This much is readily seen in a specimen of muscular fibre taken from any part of the body, and whether alive or dead. But the results of the ultimate optical analysis of these appearances, and the conclusions respecting the normal structure of striped muscle which may be legitimately drawn from them, have been the subjects of much controversy.

Quiescent muscular fibres from the chela of the forceps of a crayfish, examined while still living, without the addition of any extraneous fluid, and with magnifying powers of not less than seven or eight hundred diameters, exhibit the following appearance. At intervals of about 1-4000th of an inch, very delicate but dark and well-defined transverse lines are visible; and these, on careful focussing, appear beaded, as if they were made of a series of close-set minute granules not more than 1-20,000th to 1-30,000th of an inch in diameter. These may be termed the *septal lines* (fig. 52, D and E, *a*; C, *1—5*; fig. 53, *s*). On each side of every septal line there is a very narrow perfectly transparent band, which may be distinguished as the *septal zone* (fig. 53, *sz*). Upon this follows a relatively broad band of a substance which has a semi-transparent aspect, like very finely ground glass, and hence appears somewhat dark relatively to the septal zone. Upon this *inter-septal zone* (*i s*) follows another septal zone, then a septal line, another septal zone, an inter-septal zone, and so on throughout the whole length of the fibre.

In the perfectly unaltered state of the muscle no other transverse markings than these are discernible. But it is always possible to observe certain longitudinal markings; and these are of three kinds. In the first place, the nuclei which, in the perfectly fresh muscle, are delicate transparent oval bodies, are lodged in spaces which taper off at each end into narrow longitudinal clefts (fig. 52, A, B). Prolongations of the protoplasmic sheath of the fibre extend inwards and fill these clefts. Secondly, there are similar clefts interposed between these, but narrow and merely linear throughout. Sometimes these clefts contain fine granules. Thirdly, even in the perfectly fresh muscle, extremely faint parallel longitudinal striæ 1-7,000th of an inch, or thereabouts, apart, traverse the several zones, so that longer or shorter segments of the successive septal lines are inclosed between them. A transverse section of the muscle appears divided into rounded or polygonal areæ of the same diameter, separated from one another here and there by minute interstices. Moreover, on examination of perfectly fresh muscle with high magnifying powers, the septal lines are hardly ever straight for any distance, but are broken up into short lengths, which answer to one or more of the longitudinal divisions, and stand at slightly different heights.

The only conclusion to be drawn from these appearances seems to me to be that the substance of the muscle is composed of distinct *fibrils;* and that the longitudinal striæ and the rounded areæ of the transverse section are simply the optical expressions of the boundaries of these fibrils. In the perfectly unaltered state of the tissue, however, the fibrils are so closely packed that their boundaries are scarcely discernible.

Thus each muscular *fibre* may be regarded as composed of larger and smaller bundles of *fibrils* imbedded in a nucleated protoplasmic framework which ensheaths the whole and is itself invested by the sarcolemma.

As the fibre dies, the nuclei acquire hard, dark contours and their contents become granular, while at the same time the fibrils acquire sharp and well-defined boundaries. In fact, the fibre may now be readily teased out with needles, and the fibrils isolated.

In muscle which has been treated with various reagents, such as alcohol, nitric acid, or solution of common salt, the fibrils themselves may be split up into filaments of extreme tenuity, each of which appears to answer to one of the granules of the septal lines. Such an isolated *muscle filament* looks like a very fine thread carrying minute beads at regular intervals.

The septal lines resist most reagents, and remain visible in muscular fibres which have been subjected to various modes of treatment; but they may have the appearance of continuous bars, or be more or less completely resolved into separate granules, according to circumstances. On the other hand, what is to be seen in the interspace between every two septal lines depends upon the reagent employed. With dilute acids and strong solutions of salt, the inter-septal substance swells up and becomes transparent, so that it ceases to be distinguishable from the septal zone. At the same time a distinct but faint transverse line may appear in the middle of its length. Strong nitric acid, on the contrary, renders the inter-septal substance more opaque, and the septal zones consequently appear very well defined.

In living and recently dead muscle, as well as in muscles which have been preserved in spirit or hardened with nitric acid, the inter-septal zones polarize light; and hence, in the dark field of the polarizing microscope, the fibre appears crossed by bright bands, which correspond with the inter-septal zones, or at any rate, with the middle parts of them. The substance which forms the septal zones, on the contrary, produces no such effect, and consequently remains dark; while the septal lines again have the same property as the inter-septal substance, though in a less degree.

In fibres which have been acted upon by solution o salt, or dilute acids, the inter-septal zones have lost their polarizing property. As we know that the reagents in question dissolve the peculiar constituent of muscle, *myosin*, it is to be concluded that the inter-septal substance is chiefly composed of myosin.

Thus a fibril may be considered to be made up of

segments of different material arranged in regular order; S—sz—IS—sz—S—sz—IS—sz—S: S representing the septal line; sz, the septal zone; IS, the inter-septal zone. Of these, IS is the chief if not the only seat of the myosin; what the composition of sz and of S may be is uncertain, but the supposition, that, in the living muscle, sz is a mere fluid, appears to me to be wholly inadmissible.

When living muscle contracts, the inter-septal zones become shorter and wider and their margins darker, while the septal zones and the septal lines tend to become effaced—as it appears to me simply in consequence of the approximation of the lateral margins of the inter-septal zones. It is probable that the substance of the intermediate zone is the chief, if not the only, seat of the activity of the muscle during contraction.

5. The elements of the *nervous tissue* are of two kinds, *nerve-cells*, and *nerve fibres;* the former are found in the ganglia, and they vary very much in size (fig. 54, B). Each ganglionic corpuscle consists of a cell body produced into one or more processes which sometimes, if not always, end in nerve fibres. A large, clear spherical nucleus is seen in the interior of the nerve-cell; and in the centre of this is a well defined, small round particle, the *nucleolus.* The corpuscle, when isolated, is often surrounded by a sort of sheath of small nucleated cells.

The nerve fibres (fig. 55) of the crayfish are remarkable for the large size which some of them attain. In the central nervous system a few reach as much as 1-200th of an inch in diameter; and fibres of 1-300th or 1-400th of

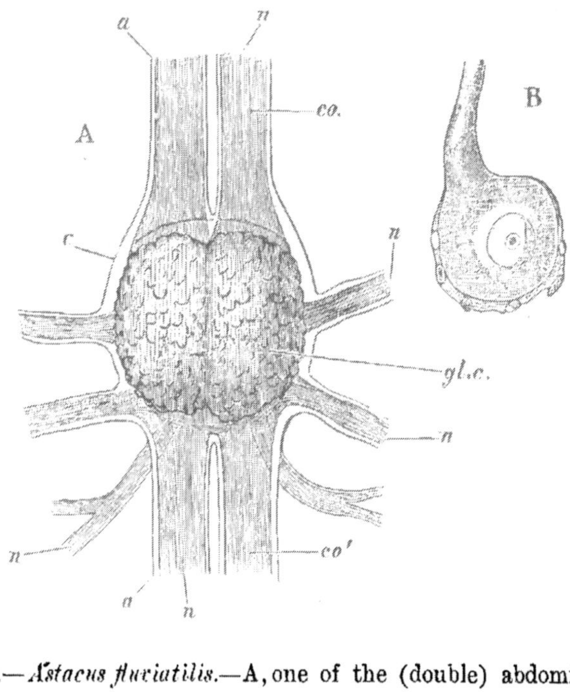

FIG. 54.—*Astacus fluviatilis.*—A, one of the (double) abdominal ganglia, with the nerves connected with it (× 25); B, a nerve cell or ganglionic corpuscle (× 250). *a*, sheath of the nerves; *c*, sheath of the ganglion; *co*, *co'*. commissural cords connecting the ganglia with those in front, and those behind them. *gl.c.* points to the ganglionic corpuscles of the ganglia; *n*, nerve fibres.

an inch in diameter are not rare in the main branches. Each fibre is a tube, formed of a strong and elastic, sometimes fibrillated, sheath, in which nuclei are imbedded at irregular intervals; and, when the nerve trunk gives off a branch, more or fewer of these tubes divide, sending off a prolongation into each branch.

When quite fresh, the contents of the tubes are perfectly pellucid, and without the least indication of structure; and, from the manner in which the contents

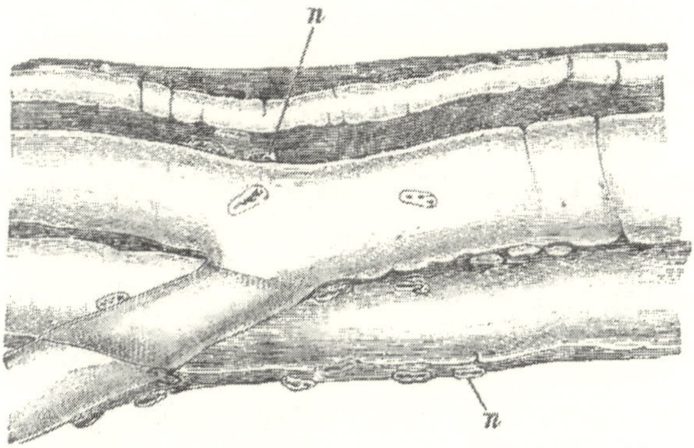

FIG. 55.—*Astacus fluviatilis.*—Three nerve fibres, with the connective tissue in which they are imbedded (magnified about 250 diameters); *n*, nuclei.

exude from the cut ends of the tubes, it is evident that they consist of a fluid of gelatinous consistency. As the fibre dies, and under the influence of water and of many chemical re-agents, the contents break up into globules or become turbid and finely granular.

Where motor nerve fibres terminate in the muscles to which they are distributed, the sheath of each fibre becomes continuous with the sarcolemma of the muscle, and the subjacent protoplasm is commonly raised into a small prominence which contains several nuclei (fig. 52, F). These are called the *terminal* or *motor plates*.

6, 7. The *ova* and the *spermatozoa* have already been described (pp. 132—135).

It will be observed that the blood corpuscles, the epithelial tissues, the ganglionic corpuscles, the ova and the spermatozoa, are all demonstrably nucleated cells, more or less modified. The first form of connective tissue is so similar to epithelial tissue, that it may obviously be regarded as an aggregate of as many cells as it presents nuclei, the matrix representing the more or less modified and confluent bodies of the cells, or products of these. But if this be so, then the second and third forms have a similar composition, except so far as the matrix of the cells has become fibrillated, or vacuolated, or marked off into masses corresponding with the several nuclei. By a parity of reasoning, muscular tissue may also be considered a cell aggregate, in which the internuclear substance has become converted into striated muscle; while, in the nerve fibres, a like process of metamorphosis may have given rise to the pellucid gelatinous nerve substance. But, if we accept the conclusions thus suggested by the comparison of the various tissues with one another, it follows that every histological element, which has now been mentioned, is either a simple nucleated cell, a modified nucleated cell, or a more or less modified cell aggregate. In other words, every tissue is resolvable into nucleated cells.

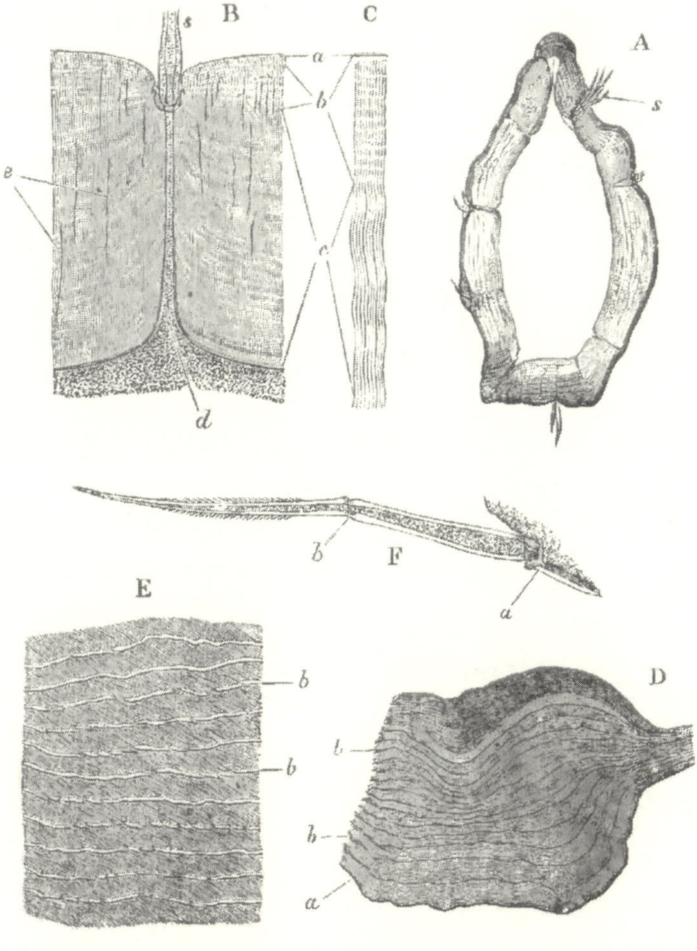

FIG. 56.—*Astacus fluviatilis.*—The structure of the cuticle. *A*, transverse section of a joint of the forceps (× 4); *s*, setæ; *B*, a portion of the same (× 30); *C*, a portion of B more highly magnified. *a*, epiostracum; *b*, ectostracum; *c*, endostracum; *d*, canal of seta; *e*, canals filled with air; *s*, seta. *D*, section of an intersternal membrane of the abdomen, the portion to the right in the natural condition, the remainder pulled apart with needles (× 20); *E*, small portion of the same, highly magnified; *a*, intermediate substance: *b*, laminæ. *F*, a seta, highly magnified; *a* and *b*, joints.

A notable exception to this generalisation, however, obtains in the case of the *cuticular structures*, in which no cellular components are discoverable. In its simplest form, such as that presented by the lining of the intestine, the cuticle is a delicate, transparent membrane, thrown off from the surface of the subjacent cells, either by a process of exudation, or by the chemical transformation of their superficial layer. No pores are discernible in this membrane, but scattered over its surface there are oval patches of extremely minute, sharp conical processes, which are rarely more than 1-5,000th of an inch long. Where the cuticle is thicker, as in the stomach and in the exoskeleton, it presents a stratified appearance, as if it were composed of a number of laminæ, of varying thickness, which had been successively thrown off from the subjacent cells.

Where the cuticular layer of the integument is uncalcified, for example, between the sterna of the abdominal somites, it presents an external, thin, dense, wrinkled lamina, the *epiostracum,* followed by a soft substance, which, on vertical section, presents numerous alternately more transparent and more opaque bands, which run parallel with one another and with the free surfaces of the slice (fig. 56, D). These bands are very close-set, often not more than 1-5000th of an inch apart near the outer and the inner surfaces, but in the middle of the section they are more distant.

If a thin vertical slice of the soft cuticle is gently

pulled with needles in the direction of its depth, it stretches to eight or ten times its previous diameter, the clear intervals between the dark bands becoming proportionally enlarged, especially in the middle of the slice, while the dark bands themselves become apparently thinner, and more sharply defined. The dark bands may then be readily drawn to a distance of as much as 1·300th of an inch from one another; but if the slice is stretched further, it splits along, or close to, one of the dark lines. The whole of the cuticular layer is stained by such colouring matters as hæmatoxylin; and, as the dark bands become more deeply coloured than the intermediate transparent substance, the transverse stratification is made very manifest by this treatment.

Examined with a high magnifying power, the transparent substance is seen to be traversed by close-set, faint, vertical lines, while the dark bands are shown to be produced by the cut edges of delicate laminæ, having a finely striated appearance, as if they were composed of delicate parallel wavy fibrillæ.

In the calcified parts of the exoskeleton a thin, tough, wrinkled epiostracum (fig. 56, B, *a*), and, subjacent to this, a number of alternately lighter and darker strata are similarly discernible: though all but the innermost laminæ are hardened by a deposit of calcareous salts, which are generally evenly diffused, but sometimes take the shape of rounded masses with irregular contours.

Immediately beneath the epiostracum there is a zone

which may occupy a sixth or a seventh of the thickness of the whole, which is more transparent than the rest, and often presents hardly any trace of horizontal or vertical striation. When it appears laminated, the strata are very thin. This zone may be distinguished as the *ectostracum* (*b*), from the *endostracum* (*c*), which makes up the rest of the exoskeleton. In the outer part of the endostracum, the strata are distinct, and may be as much as 1-500th of an inch thick, but in the inner part they become very thin, and the lines which separate them may be not more than 1-8000th of an inch apart. Fine, parallel, close-set, vertical striæ (*e*) traverse all the strata of the endostracum, and may usually be traced through the ectostracum, though they are always faint, and sometimes hardly discernible, in this region. When a high magnifying power is employed, it is seen that these striæ, which are about 1-7000th of an inch apart, are not straight, but that they present regular short undulations, the alternate convexities and concavities of which correspond with the light and the dark bands respectively.

If the hard exoskeleton has been allowed to become partially or wholly dry before the section is made, the latter will look white by reflected and black by transmitted light, in consequence of the places of the striæ being taken by threads of air of such extreme tenuity, that they may measure not more than 1-30,000th of an inch in diameter. It is to be concluded, therefore, that the striæ are the optical indications of parallel undulating canals which traverse the successive strata of the cuticle, and are ordinarily occupied by a fluid. When this dries up, the surrounding air enters, and more or less completely fills the tubes. And that this is really the case may be proved by making very thin sections parallel with the face of the exoskeleton, for these exhibit innumerable minute perforations, set at regular distances from one another, which correspond with the intervals between the striæ in the vertical section; and sometimes the contours of the areæ which separate the apertures are so well defined as to suggest a pavement of minute angular blocks, the corners of which do not quite meet.

When a portion of the hard exoskeleton is decalcified, a chitinous substance remains, which presents the same structure as that just described, except that the epiostracum is more distinct; while the ectostracum appears made up of very thin laminæ, and the tubes are represented by delicate striæ, which appear coarser in the region of the dark zones. As in the naturally soft parts of the exoskeleton, the decalcified cuticle may be split into flakes, and the pores are then seen to be disposed in distinct areæ circumscribed by clear polygonal borders. These perforated areæ appear to correspond with individual cells of the ectoderm, and the canals thus answer to the so-called "pore-canals," which are common in cuticular structures and in the walls of many cells which bound free surfaces.

The whole exoskeleton of the crayfish is, in fact, produced by the cells which underlie it, either by the exudation of a chitinous substance, which subsequently hardens, from them; or, as is more probable, by the chemical metamorphosis of a superficial zone of the bodies of the cells into chitin. However this may be, the cuticular products of adjacent cells at first form a simple, continuous, thin pellicle. A continuation of the process by which it was originated increases the thickness of the cuticle; but the material thus added to the inner surface of the latter is not always of the same nature, but is alternately denser and softer. The denser material gives rise to the tough laminæ, the softer to the intermediate transparent substance. But the quantity of the latter is at first very small, whence the more external laminæ are in close apposition. Subsequently the quantity of the intermediate substance increases, and gives rise to the thick stratification of the middle region, while it remains insignificant in the inner region of the exoskeleton.

The cuticular structures of the crayfish differ from the nails, hairs, hoofs, and similar hard parts of the higher animals, insomuch as the latter consist of aggregations of cells, the bodies of which have been metamorphosed into horny matter. The cuticle, with all its dependencies, on the contrary, though no less dependent on cells for its existence, is a derivative product, the formation of which does not involve the complete metamorphosis and consequent destruction of the cells to which it owes its origin.

The calcareous salts by which the calcified exoskeleton is hardened can only be supplied by the infiltration of a fluid in which they are dissolved from the blood; while the distinctive structural characters of the epiostracum, the ectostracum, and the endostracum, are the results of a process of metamorphosis which goes on *pari passu* with this infiltration. To what extent this metamorphosis is a properly vital process; and to what extent it is explicable by the ordinary physical and chemical properties of the animal membrane on the one hand, and the mineral salts on the other, is a curious, and at present, unsolved problem.

The outer surface of the cuticle is rarely smooth. Generally it is more or less obviously ridged or tuberculated; and, in addition, presents coarser or finer hair-like processes which exhibit every gradation from a fine microscopic down to stout spines. As these processes, though so similar to hairs in general appearance, are essentially different from the structures known as hairs in the higher animals, it is better to speak of them as *setæ*.

These setæ (fig. 56, F) are sometimes short, slender, conical filaments, the surface of which is quite smooth; sometimes the surface is produced into minute serrations, or scale-like prominences, disposed in two or more series; in other setæ, the axis gives off slender lateral branches; and in the most complicated form the branches are ornamented with lateral branchlets. For a certain distance from the base of the seta, its surface is usually smooth, even when the rest of its extent is ornamented with scales or branches. Moreover, the basal part of the seta is marked off from its apical moiety by a sort of joint which is indicated by a slight constriction, or by a peculiarity in the structure of the cuticula at this point. A seta almost always takes its origin from the bottom of a depression or pit of the layer of cuticle, from which it is developed, and at its junction with the latter it is generally thin and flexible, so that the seta moves easily in its socket. Each seta contains a cavity, the boundaries of which generally follow the outer contours of the seta. In a good many of the setæ, however, the parietes, near the base of the seta, are thickened in such a manner as almost, or completely, to obliterate the central cavity. However thick the cuticle may be at the point from which the setæ take their origin, it is always traversed by a funnel-shaped canal (fig. 56, B, *d*), which usually expands beneath the base of the seta. Through this canal the subjacent ectoderm extends up to the base of the seta, and can even be traced for some distance into its interior.

It has already been mentioned that the apodemata and the tendons of the muscles are infoldings of the cuticle, embraced and secreted by corresponding involutions of the ectoderm.

Thus the body of the crayfish is resolvable, in the first place, into a repetition of similar segments, the *metameres*, each of which consists of a somite and two appendages; the metameres are built up out of a few simple *tissues;* and, finally, the tissues are either aggregates of more or less modified nucleated *cells*, or are products of such cells. Hence, in ultimate morphological analysis, the crayfish is a multiple of the histological unit, the nucleated cell.

What is true of the crayfish, is certainly true of all animals, above the very lowest. And it cannot yet be considered certain that the generalization fails to hold good even of the simplest manifestations of animal life; since recent investigations have demonstrated the presence of a nucleus in organisms in which it had hitherto appeared to be absent.

However this may be, there is no doubt that in the case of man and of all vertebrated animals, in that of all arthropods, mollusks, echinoderms, worms, and inferior organisms down to the very lowest sponges, the process of morphological analysis yields the same result as in the case of the crayfish. The body is built up of tissues, and the tissues are either obviously composed of nucleated cells; or, from the presence of nuclei, they may be assumed to be the results of the metamorphosis of such cells; or they are cuticular structures.

The essential character of the nucleated cell is that it consists of a protoplasmic substance, one part of which differs somewhat in its physical and chemical characters from the rest, and constitutes the nucleus. What part the nucleus plays in relation to the functions, or vital activities, of the cell is as yet unknown; but that it is the seat of operations of a different character from those which go on in the body of the cell is clear enough. For, as we have seen, however different the several tissues may be, the nuclei which they contain are very much alike; whence it follows, that if all these tissues were primitively composed of simple nucleated cells, it must be the bodies of the cells which have undergone metamorphosis, while the nuclei have remained relatively unchanged.

On the other hand, when cells multiply, as they do in all growing parts, by the division of one cell into two, the signs of the process of internal change which ends in fission are apparent in the nucleus before they are manifest in the body of the cell; and, commonly, the division of the former precedes that of the latter. Thus a single cell body may possess two nuclei, and may become divided into two cells by the subsequent aggregation of the two moieties of its protoplasmic substance round each of them, as a centre.

In some cases, very singular structural changes take place in the nuclei in the course of the process of cell-division. The granular or fibrillar contents of the nucleus, the wall of which becomes less distinct, arrange themselves in the form of a spindle or double cone, formed of extremely delicate filaments; and in the plane of the base of the double cone the filaments present knots or thickenings, just as if they were so many threads with a bead in the middle of each. When the nuclear spindle is viewed sideways, these beads or thickenings give rise to the appearance of a disk traversing the centre of the spindle. Soon each bead separates into two, and these move away from one another, but remain connected by a fine filament. Thus the structure which had the form of a double cone, with a disk in the middle, assumes that of a short cylinder, with a disk and a cone at each end. But as the distance between the two disks increases, the uniting filaments lose their parallelism, converge in the middle, and finally separate, so that two separate double cones are developed in place of the single one. Along with these changes in the nucleus, others occur in the protoplasm of the cell body, and its parts commonly display a tendency to arrange themselves in radii from the extremities of the cones as a centre; while, as the separation of the two secondary nuclear spindles becomes complete, the cell body gradually splits from the periphery inwards, in a direction at right angles to the common axis of the spindles and between their apices. Thus two cells are formed, where, previously, only one existed; and the nuclear spindles of each soon revert to the globular form and confused arrangement of the contents, characteristic of nuclei in their ordinary state. The formation of these nuclear spindles is very beautifully seen in the epithelial cells of the testis of the crayfish (fig. 33, p. 132); but I have not been able to find distinct evidence of it elsewhere in this animal; and although the process has now been proved to take place in all the divisions of the animal kingdom, it would seem that nuclei may, and largely do, undergo division, without becoming converted into spindles.

The most cursory examination of any of the higher plants shows that the vegetable, like the animal body, is made up of various kinds of tissues, such as pith, woody fibre, spiral vessels, ducts, and so on. But even the most modified forms of vegetable tissue depart so little from the type of the simple cell, that the reduction of them all to that common type is suggested still more strongly than in the case of the animal fabric. And thus the nucleated cell appears to be the morphological unit of the plant no less than of the animal. Moreover, recent inquiry has shown that in the course of the multiplication of vegetable cells by division, the nuclear spindles may appear and run through all their remarkable changes by stages precisely similar to those which occur in animals.

The question of the universal presence of nuclei in cells may be left open in the case of Plants, as in that of Animals; but, speaking generally, it may justly be affirmed that the nucleated cell is the morphological foundation of both divisions of the living world; and the great generalisation of Schleiden and Schwann, that there is a fundamental agreement in structure and development between plants and animals, has, in substance, been merely confirmed and illustrated by the labours of the half century which has elapsed since its promulgation.

Not only is it true that the minute structure of the crayfish is, in principle, the same as that of any other animal, or of any plant, however different it may be in detail; but, in all animals (save some exceptional forms) above the lowest, the body is similarly composed of three layers, ectoderm, mesoderm, and endoderm, disposed around a central alimentary cavity. The ectoderm and the endoderm always retain their epithelial character; while the mesoderm, which is insignificant in the lower organisms, becomes, in the higher, far more complicated even than it is in the crayfish.

Moreover, in the whole of the *Arthropoda,* and the whole of the *Vertebrata,* to say nothing of other groups of animals, the body, as in the crayfish, is susceptible of distinction into a series of more or less numerous segments, composed of homologous parts. In each segment these parts are modified according to physiological requirements; and by the coalescence, segregation, and change of relative size and position of the segments, well characterized regions of the body are marked out. And it is remarkable that precisely the same principles are illustrated by the morphology of plants. A flower with its whorls of sepals, petals, stamens and carpels has the same relation to a stem with its whorls of leaves, as a crayfish's head has to its abdomen, or a dog's skull to its thorax.

It may be objected, however, that the morphological generalisations which have now been reached, are to a considerable extent of a speculative character; and that, in the case of our crayfish, the facts warrant no more than the assertion that the structure of that animal may be consistently interpreted, on the supposition that the body is made up of homologous somites and appendages, and that the tissues are the result of the modification of homologous histological elements or cells; and the objection is perfectly valid.

There can be no doubt that blood corpuscles, liver cells, and ova are all nucleated cells; nor any that the third, fourth, and fifth somites of the abdomen are constructed upon the same plan; for these propositions are mere statements of the anatomical facts. But when, from the presence of nuclei in connective tissue and muscles, we conclude that these tissues are composed of modified cells; or when we say that the ambulatory limbs of the thorax are of the same type as the abdominal limbs, the exopodite being suppressed, the statement, as the evidence stands at present, is no more than a convenient way of interpreting the facts. The question remains, has the muscle actually been formed out of nucleated cells? Has the ambulatory limb ever possessed an exopodite, and lost it?

The answer to these questions is to be sought in the facts of individual and ancestral development.

An animal not only is, but becomes; the crayfish is the product of an egg, in which not a single structure visible in the adult animal exists: in that egg the different tissues and organs make their appearance by a gradual process of evolution; and the study of this process can alone tell us whether the unity of composition suggested by the comparison of adult structures, is borne out by the facts of their development in the individual or not. The hypothesis that the body of the crayfish is made up of a series of homologous somites and appendages, and that all the tissues are composed of nucleated cells, might be only a permissible, because a useful, mode of colligating the facts of anatomy. The investigation of the actual manner in which the evolution of the body of the crayfish has been effected, is the only means of ascertaining whether it is anything more. And, in this sense, development is the criterion of all morphological speculations.

The first obvious change which takes place in an impregnated ovum is the breaking up of the yelk into smaller portions, each of which is provided with a nucleus, and is termed a *blastomere*. In a general morphological sense, a blastomere is a nucleated cell, and differs from an ordinary cell only in size, and in the usual, though by no means invariable, abundance of granular contents; and blastomeres insensibly pass into ordinary cells, as the process of division of the yelk into smaller and smaller portions goes on.

In a great many animals, the splitting-up into blastomeres is effected in such a manner that the yelk is, at first, divided into equal, or nearly equal, masses; that each of these again divides into two; and that the number of blastomeres thus increases in geometrical progression until the entire yelk is converted into a mulberry-like body, termed a *morula*, made up of a great number of small blastomeres or nucleated cells. The whole organism is subsequently built up by the multiplication, the change of position, and the metamorphosis of these products of yelk division.

In such a case as this, yelk division is said to be *complete*. An unessential modification of complete yelk division is seen when, at an early period, the blastomeres produced by division are of unequal sizes; or when they become unequal in consequence of division taking place much more rapidly in one set than in another.

In many animals, especially those which have large ova, the inequality of division is pushed so far that only a portion of the yelk is affected by the process of fission, while the rest serves merely as *food-yelk*, for nutriment to the blastomeres thus produced. Over a greater or less extent of the surface of the egg, the protoplasmic substance of the yelk segregates itself from the rest, and, constituting a *germinal layer*, breaks up into the blastomeres, which multiply at the expense of the food-yelk, and fabricate the body of the embryo. This process is termed *partial* or *incomplete* yelk division.

The crayfish is one of those animals in the egg of which the yelk undergoes partial division. The first steps of the process have not yet been thoroughly worked out, but their result is seen in ova which have been but a short time laid (fig. 57, A). In such eggs, the great mass of the substance of the vitellus is destined to play the part of food-yelk; and it is disposed in conical masses, which radiate from a central spheroidal portion to the periphery of the yelk (*v*). Corresponding with the base of each cone, there is a clear protoplasmic plate, which contains a nucleus; and as these bodies are all in contact by their edges, they form a complete, though thin, investment to the food-yelk. This is termed the *blastoderm* (*bl*).

Each nucleated protoplasmic plate adheres firmly to the corresponding cone of granular food-yelk, and, in all probability, the two together represent a blastomere; but, as the cones only indirectly subserve the growth of the embryo, while the nucleated peripheral plates form an independent spherical sac, out of which the body of the young crayfish is gradually fashioned, it will be convenient to deal with the latter separately.

Thus, at this period, the body of the developing crayfish is nothing but a spherical bag, the thin walls of which are composed of a single layer of nucleated cells, while its cavity is filled with food-yelk. The first modification

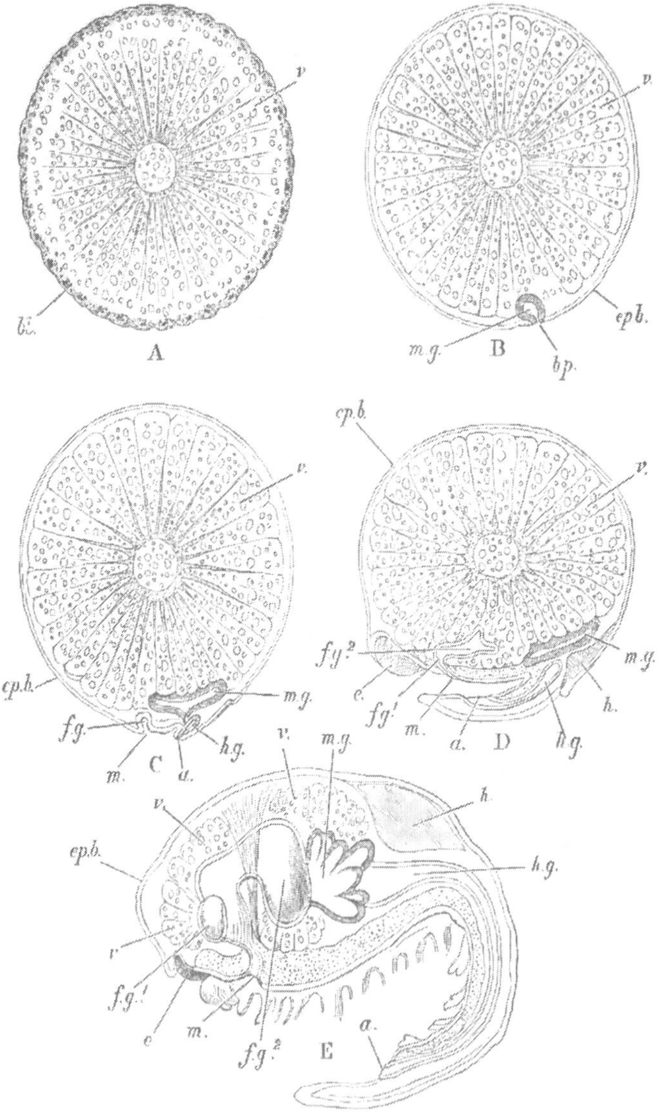

FIG. 57.—*Astacus fluviatilis.*—Diagrammatic sections of embryos; partly after Reichenbach, partly original (×20). A. An ovum in which the blastoderm is just formed. B. An ovum in which the invagination of the blastoderm to constitute the hypoblast or rudiment of the mid-gut has taken place. (This nearly answers to the stage represented in fig. 58, *A.*) C. A longitudinal section of an ovum, in which the rudiments of the abdomen, of the hind-gut, and of the fore-gut have appeared. (This nearly answers to the stage represented in fig. 58, E.) D. A similar section of an embryo in nearly the same stage of development as that represented in C, fig. 59. E. An embryo just hatched, in longitudinal section; *a*, anus; *bl*, blastoderm; *bp*, blastopore; *e*, eye; *ep.b.*, epiblast; *fg*, fore-gut; fg^1, its œsophageal, and fg^2, its gastric portion; *h*, heart; *hg*, hind-gut; *m*, mouth; *mg*, hypoblast, archenteron, or mid-gut; *v*, yelk. The dotted portions in D and E represent the nervous system.

which is effected in the vesicular blastoderm manifests itself on that face of it which is turned towards the pedicle of the egg. Here the layer of cells becomes thickened throughout an oval area about 1-25th of an inch in diameter. Hence, when the egg is viewed by reflected light, a whitish patch of corresponding form and size appears in this region. This may be termed the *germinal disk.* Its long axis corresponds with that of the future crayfish.

Next, a depression (fig. 58, A, *bp*) appears in the hinder third of the germinal disk, in consequence of this part of the blastoderm growing inwards, and thus giving rise to a small wide-mouthed pouch, which projects into the food-yelk with which the cavity of the blastoderm is filled (fig. 57, B, *mg*). As this infolding, or invagination of the blastoderm, goes on, the pouch thus produced increases, while its external opening, termed the *blastopore* (fig. 57, B, and 58, A—E, *bp*), diminishes in size. Thus the body of the embryo crayfish, from being a simple bag becomes a double bag, such as might be produced by pushing in the wall of an incompletely distended india-rubber ball with the finger. And, in this case, if the interior of the bag contained porridge, the latter would very fairly represent the food-yelk.

By this invagination a most important step has been taken in the development of the crayfish. For, though the pouch is nothing but an ingrowth of part of the blastoderm, the cells of which its wall is composed

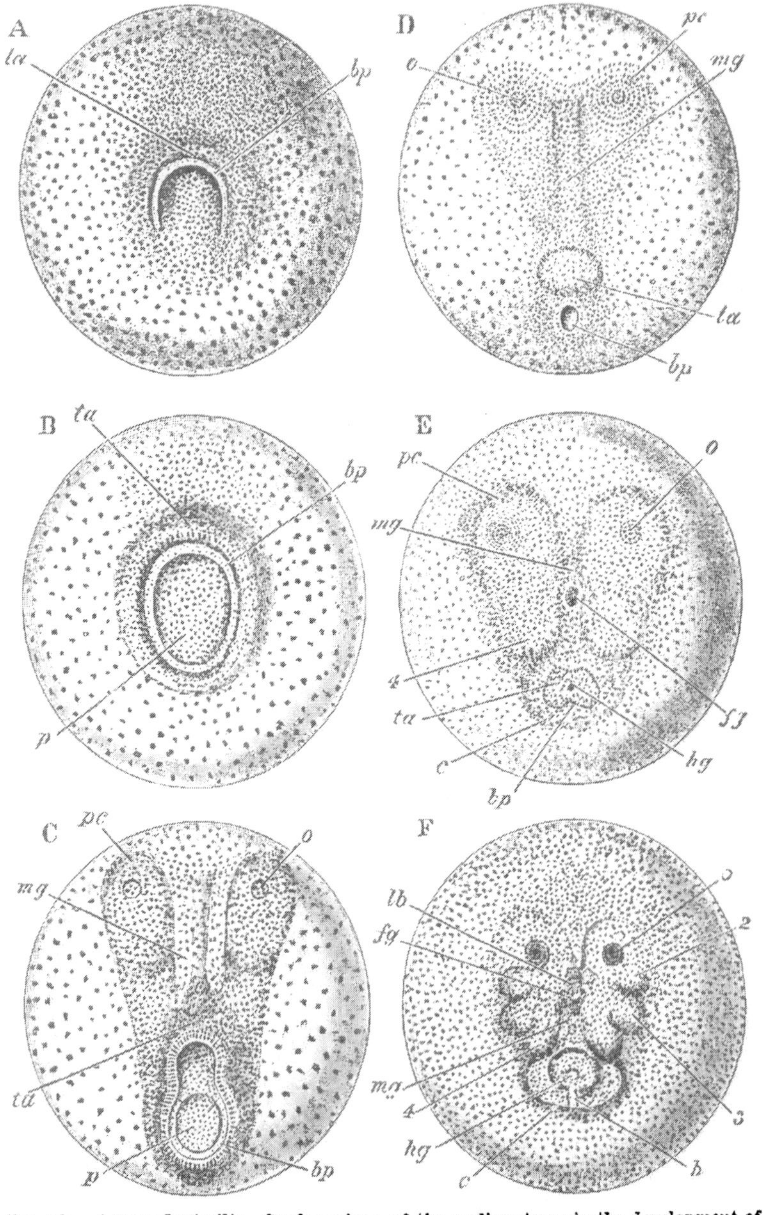

FIG. 58.—*Astacus fluviatilis.*—Surface views of the earlier stages in the development of the embryo, from the appearance of the blastopore (A) to the assumption of the nauplius form (F) (after Reichenbach, × about 23). *bp*, blastopore, *c*, carapace; *fg*, fore-gut involution; *h*, heart; *hg*, hind-gut involution; *lb*, labrum; *mg*, medullary groove; *o*, optic pit; *p*, endodermal plug partly filling up the blastopore; *pc*, procephalic processes; *ta*, abdominal elevation; *2*, antennules; *3*, antennæ *4*, mandibles.

henceforward exhibit different tendencies from those which are possessed by the rest of the blastoderm. In fact, it is the primitive alimentary apparatus or *archenteron*, and its wall is termed the *hypoblast*. The rest of the blastoderm, on the contrary, is the primitive epidermis, and receives the name of *epiblast*. If the food-yelk were away, and the archenteron enlarged until the hypoblast came in contact with the epiblast, the entire body would be a double-walled sac, containing an alimentary cavity, with a single external aperture. This is the *gastrula* condition of the embryo; and some animals, such as the common fresh-water polype, are little more than permanent *gastrulæ*.

Although the gastrula has not the slightest resemblance to a crayfish, yet, as soon as the hypoblast and the epiblast are thus differentiated, the foundations of some of the most important systems of organs of the future crustacean are laid. The hypoblast will give rise to the epithelial lining of the mid-gut; the epiblast (which answers to the ectoderm in the adult) to the epithelia of the fore-gut and hind-gut, to the epidermis, and to the central nervous system.

The mesodermal structures, that is to say the connective tissue, the muscles, the heart and vessels, and the reproductive organs, which lie between the ectoderm and the endoderm, are not derived directly from either the epiblast or the hypoblast, but have a *quasi*-independent origin, from a mass of cells which first makes its appearance in the neighbourhood of the blastopore, between the hypoblast and the epiblast, though they are probably derived from the former. From this region they gradually spread, first over the sternal, and then on to the tergal aspect of the embryo, and constitute the *mesoblast.*

Epiblast, hypoblast, and mesoblast are at first alike constituted of nothing but nucleated cells, and they increase in dimensions by the continual fission and growth of these cells. The several layers become gradually modelled into the organs which they constitute, before the cells undergo any notable modification into tissues. A limb, for example, is, at first, a mere cellular outgrowth, or bud, composed of an outer coat of epiblast with an inner core of mesoblast; and it is only subsequently that its component cells are metamorphosed into well-defined epidermic and connective tissues, vessels and muscles.

The embryo crayfish remains only a short while in the gastrula stage, as the blastopore soon closes up, and the archenteron takes the form of a sac, flattened out between the epiblast and the food-yelk, with which its cells are in close contact (fig. 57, C and D).* Indeed, as development proceeds, the cells of the hypoblast actually feed upon the substance of the food-yelk, and turn it to account for the general nutrition of the body.

* Whether, as some observers state, the hypoblastic cells grow over and inclose the food-yelk or not, is a question that may be left open. I have not been able to satisfy myself of this fact.

The sternal area of the embryo gradually enlarges until it occupies one hemisphere of the yelk; in other words, the thickening of the epiblast gradually extends outwards. Just in front of the blastopore, as it closes, the middle of the epiblast grows out into a rounded elevation (fig. 58, *t a*; fig. 59, *ab*), which rapidly increases in length, and at the same time turns forwards. This is the rudiment of the whole abdomen of the crayfish. Further forwards, two broad and elongated, but flatter thickenings appear; one on each side of the middle line (fig. 58, *p c*). As the free end of the abdominal papilla now marks the hinder extremity of the embryo, so do these two elevations, which are termed the *procephalic lobes*, define its anterior termination. The whole sternal region of the body will be produced by the elongation of that part of the embryo which lies between these two limits.

A narrow longitudinal groove-like depression appears on the surface of the epiblast, in the middle line, between the procephalic lobes and the base of the abdominal papilla (fig. 58, C—F, *m g*). About its centre, this groove becomes further depressed by the ingrowth of the epiblast, which constitutes its floor, and gives rise to a short tubular sac, which is the rudiment of the whole fore-gut (fig. 57, C, and fig. 58, E, *f g*). At first, this epiblastic ingrowth does not communicate with the archenteron, but, after a while, its blind end combines with the front and lower part of the hypoblast, and an opening is formed by which the cavity of the fore-gut communicates with that of the mid-gut (fig. 57, E). Thus a gullet and stomach, or rather the parts which will eventually give rise to all these, are constituted. And it is important to remark that, in comparison with the mid-gut, they are, at first, very small.

In the same way, the epiblast covering the sternal face of the abdominal papilla undergoes invagination and is converted into a narrow tube which is the origin of the whole hind-gut (fig. 57, C, and fig. 58, E, *hg*). This, like the fore-gut, is at first blind; but the shut front end soon applying itself to the hinder wall of the archenteric sac, the two coalesce and open into one another (fig. 57, E). Thus the complete alimentary canal, consisting of a very narrow, tubular, fore- and hind-gut, derived from the epiblast, and a wider and more sac-like mid-gut, formed of the whole hypoblast, is constituted.

The procephalic lobes become more convex; while, behind them, the surface of the epiblast rises into six elevations disposed in pairs, one on each side of the median groove. The hindermost of these, which lie at the sides of the mouth, are the rudiments of the mandibles (fig. 58, E and F, *4*); the other two become the antennæ (*3*) and the antennules (*2*), while, at a later period, processes of the procephalic lobes give rise to the eyestalks.

A short distance behind the abdomen, the epiblast rises into a transverse ridge, which is concave forwards, while its ends are prolonged on each side nearly as far as the mouth. This is the commencement of the free edge of the carapace (fig. 58, E and F, and fig. 59, A, *c*) —the lateral parts of which, greatly enlarging, become the branchiostegites (fig. 59, D, *c*).

In many animals allied to crayfish, the young, when it has reached a stage in its development, which answers to this, undergoes rapid changes of outward form and of internal structure, without making any essential addition to the number of the appendages. The appendages which represent the antennules, the antennæ, and the mandibles elongate and become oar-like locomotive organs; a single median eye is developed, and the young leaves the egg as an active larva, which is known as a *Nauplius*. The crayfish, on the other hand, is wholly incapable of an independent existence at this stage, and continues its embryonic life within the egg case; but it is a remarkable circumstance that the cells of the epiblast secrete a delicate cuticula, which is subsequently shed. It is as if the animal symbolized a nauplius condition by the development of this cuticle, as the fœtal whalebone whale symbolizes a toothed condition by developing teeth which are subsequently lost and never perform any function.

In fact, in the crayfish, the nauplius condition is soon left behind. The sternal disk spreads more and more over the yelk; as the region between the mouth and the root of the abdomen elongates, slight transverse

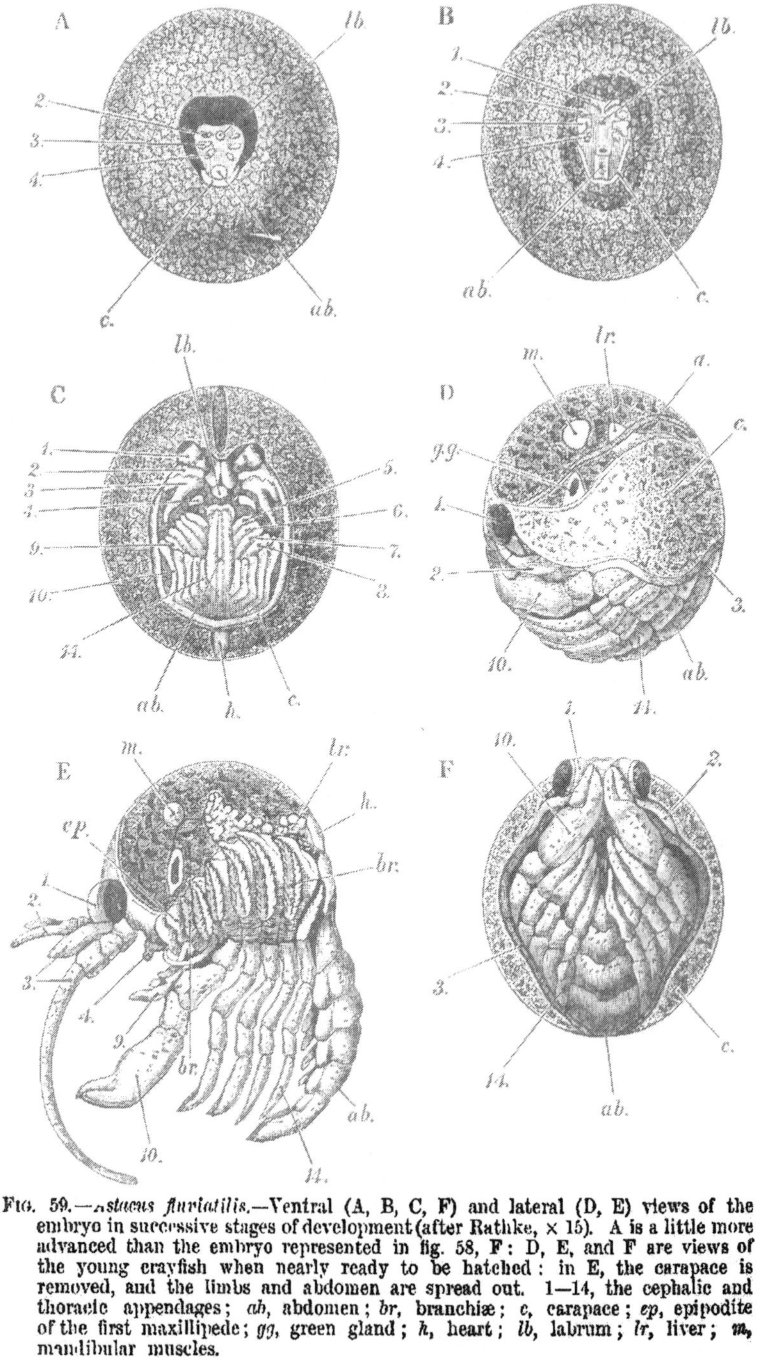

FIG. 59.—*Astacus fluviatilis.*—Ventral (A, B, C, F) and lateral (D, E) views of the embryo in successive stages of development (after Rathke, × 15). A is a little more advanced than the embryo represented in fig. 58, F: D, E, and F are views of the young crayfish when nearly ready to be hatched: in E, the carapace is removed, and the limbs and abdomen are spread out. 1—14, the cephalic and thoracic appendages; *ab*, abdomen; *br*, branchiæ; *c*, carapace; *ep*, epipodite of the first maxillipede; *gg*, green gland; *h*, heart; *lb*, labrum; *lr*, liver; *m*, mandibular muscles.

depressions indicate the boundaries of the posterior cephalic and the thoracic somites; and pairs of elevations, similar to the rudiments of the antennules and antennæ, appear upon them in regular order from before backwards (fig. 59, C).

In the meanwhile, the extremity of the abdomen flattens out and takes on the form of an oval plate, the middle of the posterior margin of which is slightly truncated or notched; while, finally, transverse constrictions mark off six segments, the somites of the abdomen, in front ot this. Along with these changes, four pairs of tubercles grow out from the sternal faces of the four middle abdominal somites, and constitute the rudiments of the four middle pairs of abdominal appendages. The first abdominal somite exhibits only two hardly perceptible elevations in place of the appendages of the others, while the sixth seems, at first, to have none. The appendages of the sixth somite, however, are already formed, though, singularly enough, they lie beneath the cuticle of the telson and are set free only after the first ecdysis.

The rostrum grows out between the procephalic lobes; it remains relatively very short up to the time that the young crayfish quits the egg, and is directed more downwards than forwards. The lateral portions of the carapacial ridge, becoming deeper, are converted into the branchiostegites, and the cavities which they overarch are the branchial chambers. The transverse portion of the ridge, on the other hand, remains relatively short, and constitutes the free posterior margin of the carapace.

As these changes take place, the abdomen and the sternal region of the thorax are constantly enlarging in proportion to the rest of the ovum; and the food-yelk which lies in the cephalothorax is, *pari passu*, being diminished. Hence the cephalothorax constantly becomes relatively smaller and the tergal aspect of the carapace less spherical; although, even when the young crayfish is ready to be hatched, the difference between it and the adult in the form of the cephalothoracic region, and in the size of the latter relatively to the abdomen, is very marked.

The simple bud-like outgrowths of the somites, in which all the appendages take their origin, are rapidly metamorphosed. The eyestalks (fig. 59, *1*) soon attain a considerable relative size. The extremities of the antennules (*2*) and of the antennæ (*3*) become bifurcated; and the two divisions of the antennule remain broad, thick, and of nearly the same size up to birth. On the other hand, the inner or endopoditic division of the antenna becomes immensely lengthened, and at the same time annulated, while the outer or exopoditic division remains relatively short, and acquires its characteristic scale-like form.

The labrum (*lb*) arises as a prolongation of the middle sternal region in front of the mouth, while the bilobed metastoma is an outgrowth of the sternal region behind it.

The posterior cephalic and the thoracic appendages (*5—14*) elongate and gradually approach the form which they possess in the adult. I have not been able to discover, at any period of development, an outer division or exopodite in any of the five posterior thoracic limbs. And this is a very remarkable circumstance, inasmuch as such an exopodite exists in the closely allied lobster in the larval state; and, in many of the shrimp and prawn-like allies of the crayfish, a complete or rudimentary exopodite is found in these limbs, even in the adult condition.

When the crayfish is hatched (fig. 60) it differs from the adult in many ways—not only is the cephalothorax more convex and larger in proportion to the abdomen; but the rostrum is short and bent down between the eyes. The sterna of the thorax are wider relatively, and hence there is a greater interval between the bases of the legs than in the adult. The proportion of the limbs to one another and to the body are nearly the same as in the adult, but the chelæ of the forceps are more slender. The tips of the chelæ are all strongly incurved (fig. 8, B, p. 41), and the dactylopodites of the two posterior thoracic limbs are hook-like. The appendages of the first abdominal somite are undeveloped, and those of the last are inclosed within the telson, which is, as has already been said, of a broad oval form, usually notched in the middle of its hinder margin, and devoid of any indication of transverse division. Its margins are produced into a single series of short conical processes, and the disposition of the vascular canals in its interior gives it the appearance of being radially striated.

The setæ, so abundant in the adult, are very scanty in the newly hatched young; and the great majority of those which exist are simple conical prolongations of the un-

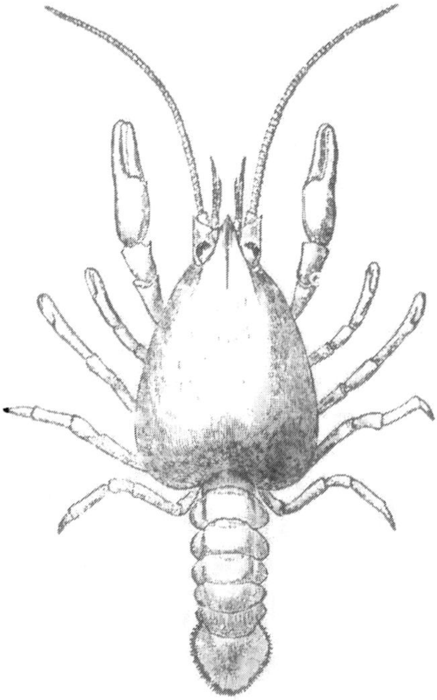

FIG. 60.—*Astacus fluviatilis.*—Newly-hatched young (× 6).

calcified cuticle, the bases of which are not sunk in pits and which are devoid of lateral scales or processes.

The young animals are firmly attached to the abdominal appendages of the parent in the manner already described. They are very sluggish, though they move when touched; and at this period they do not feed, but are nourished by the food-yelk, of which a considerable store still remains in the cephalothorax.

I imagine that they are set free during the first ecdysis, and that the appendages of the sixth abdominal somite are at that time expanded, but nothing is definitely known at present of these changes.

The foregoing sketch of the general nature of the changes which take place in the egg of the crayfish suffice to show that its development is, in the strictest sense of the word, a process of evolution. The egg is a relatively homogeneous mass of living protoplasmic matter, containing much nutritive material; and the development of the crayfish means the gradual conversion of this comparatively simple body into an organism of great complexity. The yelk becomes differentiated into formative and nutritive portions. The formative portion is subdivided into histological units: these arrange themselves into a blastodermic vesicle; the blastoderm becomes differentiated into epiblast, hypoblast, and mesoblast; and the simple vesicle assumes the gastrula condition. The layers of the gastrula shape themselves into the body of the crayfish and its appendages, while along with this, the cells of which all the parts are built, become metamorphosed into tissues, each with its characteristic properties. And all these wonderful changes are the necessary consequences of the interaction of the molecular forces resident in the substance of the impregnated ovum, with the conditions to which it is exposed; just as the forms evolved from a crystallising fluid are dependent upon the chemical composition of the dissolved matter and the influence of surrounding conditions.

Without entering into details which lie beyond the scope of the present work, something must be said respecting the manner in which the complicated internal organisation of the crayfish is evolved from the cellular double sac of the gastrula stage.

It has been seen that the fore-gut is at first an insignificant tubular involution of the epiblast in the region of the mouth. It is, in fact, a part of the epiblast turned inwards, and the cells of which it is composed secrete a thin cuticular layer, as do those of the rest of the epiblast, which gives rise to the ectodermal or epidermic part of the integument. As the embryo grows, the fore-gut enlarges much faster than the mid-gut, increasing in height and from before backwards, while its side-walls remain parallel, and are separated by only a narrow cavity. At length, it takes on the shape of a triangular bag (fig. 57, D, *fg*), attached by its narrow end around the mouth and immersed in the food-yelk, which it gradually divides into two lobes, one on the right and one on the left side. At the same time a vertical plate of mesoblastic tissue, from which the great anterior and posterior muscles are eventually developed, connects it with the roof and with the front wall of the carapace.

Becoming constricted in the middle, the fore-gut next appears to consist of two dilatations of about equal size, connected by a narrower passage (fig. 57, E, *fg*[1], *fg*[2]). The front dilatation becomes the œsophagus and the cardiac division of the stomach; the hinder one, the pyloric division. At the sides of the front end of the cardiac division two small pouches are formed shortly after birth; in each of these a thick laminated deposit of chitin takes place, and constitutes a minute crab's-eye or gastrolith, which has the same structure as in the adult, and is largely calcified. This fact is the more remarkable as, at this time, the exoskeleton contains very little calcareous deposit. In the position of the gastric teeth, folds of the cellular wall of corresponding shape are formed, and the chitinous cuticle of which the teeth are composed is, as it were, modelled upon them.

The hind-gut occupies the whole length of the abdomen, and its cells early arrange themselves into six ridges, and secrete a cuticular layer.

The mid-gut, or hypoblastic sac, very soon gives off numerous small prolongations on each side of its hinder extremity, and these are converted into the cæca of the liver (fig. 57, E, *mg*). The cells of its tergal wall are in close contact with the adjacent masses of food-yelk; and it is probable that the gradual absorption of the food-yelk is chiefly effected by these cells. At birth, however, the lateral lobes of the food-yelk are still large, and occupy the space left between the stomach and liver on the one hand, and the cephalic integument on the other.

The mesoblastic cells give rise to the layer of connective tissue which forms the deeper portion of the integument, and to that which invests the alimentary canal; to all the muscles; and to the heart, the vessels, and the corpuscles of the blood. The heart appears very early as a solid mass of mesoblastic cells in the tergal region of the thorax, just in front of the origin of the abdomen (figs. 57, 58, 59, *h*). It soon becomes hollow, and its walls exhibit rhythmical contractions.

The branchiæ are, at first, simple papillæ of the integument of the region from which they take their rise. These papillæ elongate into stems, which give off lateral filaments. The podobranchiæ are at first similar to the arthrobranchiæ, but an outgrowth soon takes place near the free end of the stem, and becomes the lamina, while the attached end enlarges into the base.

The renal organ is stated to arise by a tubular involution of the epiblast, which soon becomes convoluted, and gives rise to the green gland.

The central nervous system is wholly a product of the epiblast. The cells which lie at the sides of the longitudinal groove already mentioned (fig. 58, *mg*), grow inwards, and give rise to two cords which are at first separate from one another and continuous with the rest of the epiblast. At the front end of the groove a depression arises, and its cells form a mass which connects these two cords in front of the mouth, and gives rise to the cerebral ganglia. The epiblastic linings of two small pits (fig. 58, *o*) which appear very early on the surface of the procephalic lobes, are also carried inwards in the same way, and, uniting with the foregoing, produce the optic ganglia.

The cells of the longitudinal cords become differentiated into nerve fibres and nerve cells, and the latter, gathering towards certain points, give rise to the ganglia which eventually unite in the middle line. By degrees, the ingrowth of epiblastic cells, from which all these structures are developed, becomes completely separated from the rest of the epiblast, and is invested by mesoblastic cells. The central nervous system, therefore, in a crayfish, as in a vertebrated animal, is at first, as a part of the ectoderm, morphologically one with the epidermis; and the deep and protected position which it occupies in the adult is only a consequence of the mode in which the nervous portion of the ectoderm grows inwards and becomes detached from the epidermic portion.

The visual rods of the eye are merely modified cells of the ectoderm. The auditory sac is formed by an involution of the ectoderm of the basal joint of the antennule. At birth it is a shallow wide-mouthed depression, and contains no otoliths.

Lastly, the reproductive organs result from the segregation and special modification of cells of the mesoblast behind the liver. Rathke states that the sexual apertures are not visible until the young crayfish has attained the length of an inch; and that the first pair of abdominal appendages of the male appear still later in the form of two papillæ, which gradually elongate and take on their characteristic forms.

CHAPTER V.

THE COMPARATIVE MORPHOLOGY OF THE CRAYFISH.—THE STRUCTURE AND THE DEVELOPMENT OF THE CRAYFISH COMPARED WITH THOSE OF OTHER LIVING BEINGS.

Up to this point, our attention has been directed almost exclusively to the common English crayfish. Except in so far as the crayfish is dependent for its maintenance upon other animals, or upon plants, we might have ignored the existence of all living things except crayfishes. But, it is hardly necessary to observe, that innumerable hosts of other forms of life not only tenant the waters and the dry land, but throng the air; and that all the crayfishes in the world constitute a hardly appreciable fraction of its total living population.

Common observation leads us to see that these multitudinous living beings differ from not-living things in many ways; and when the analysis of these differences is pushed as far as we are at present able to carry it, it shews us that all living beings agree with the crayfish and differ from not-living things in the same particulars. Like the crayfish, they are constantly wasting away by oxidation, and repairing themselves by taking into their substance the matters which serve them for food; like the crayfish, they shape themselves according to a definite pattern of external form and internal structure; like the crayfish, they give off germs which grow and develope into the shapes characteristic of the adult. No mineral matter is maintained in this fashion; nor grows in the same way; nor undergoes this kind of development; nor multiplies its kind by any such process of reproduction.

Again, common observation early leads to the discrimination of living things into two great divisions. Nobody confounds ordinary animals with ordinary plants, nor doubts that the crayfish belongs to the former category and the waterweed to the latter. If a living thing moves and possesses a digestive receptacle, it is held to be an animal; if it is motionless and draws its nourishment directly from the substances which are in contact with its outer surface, it is held to be a plant. We need not inquire, at present, how far this rough definition of the differences which separate animals from plants holds good. Accepting it for the moment, it is obvious that the crayfish is unquestionably an animal,—as much an animal as the vole, the perch, and the pond-snail, which inhabit the same waters. Moreover, the crayfish has, in common with these animals, not merely the motor and digestive powers characteristic of animality, but they all, like it, possess a complete alimentary canal; special apparatus for the circulation and the aëration of the blood; a nervous system with sense-organs; muscles and motor mechanisms; reproductive organs. Regarded as pieces of physiological apparatus, there is a striking similarity between all three. But, as has already been hinted in the preceding chapter, if we look at them from a purely morphological point of view, the differences between the crayfish, the perch, and the pond-snail, appear at first sight so great, that it may be difficult to imagine that the plan of structure of the first can have any relation to that of either of the last two. On the other hand, if the crayfish is compared with the water-beetle, notwithstanding wide differences, many points of similarity between the two will manifest themselves; while, if a small lobster is set side by side with a crayfish, an unpractised observer, though he will readily see that the two animals are somewhat different, may be a long time in making out the exact nature of the differences.

Thus there are degrees of likeness and unlikeness among animals, in respect of their outward form and internal structure, or, in other words, in their morphology. The lobster is very like a crayfish, the beetle is remotely like one; the pond-snail and the perch are extremely unlike crayfishes. Facts of this kind are commonly expressed in the language of zoologists, by saying that the lobster and the crayfish are closely allied forms; that the beetle and the crayfish present a remote affinity; and that there is no affinity between the crayfish and the pond-snail, or the crayfish and the perch.

The exact determination of the resemblances and differences of animal forms by the comparison of the structure and the development of one with those of another, is the business of comparative morphology. Morphological comparison, fully and thoroughly worked out, furnishes us with the means of estimating the position of any one animal in relation to all the rest; while it shews us with what forms that animal is nearly, and with what it is remotely, allied: applied to all animals, it furnishes us with a kind of map, upon which animals are arranged in the order of their respective affinities; or a classification, in which they are grouped in that order. For the purpose of developing the results of comparative morphology in the case of the crayfish, it will be convenient to bring together, in a summary form, those points of form and structure, many of which have already been referred to and which characterise it as a separate kind of animal.

Full-grown English crayfishes usually measure about three inches and a half from the extremity of the rostrum in front to that of the telson behind. The largest specimen I have met with measured four inches.* The

* The dimensions of crayfishes at successive ages given at p. 31, beginning at the words "By the end of the year," refer to the "écrevisse à pieds rouges" of France; not to the English crayfish, which is

males are commonly somewhat larger, and they almost always have longer and stronger forceps than the females. The general colour of the integument varies from a light reddish-brown to a dark olive-green; and the hue of the tergal surface of the body and limbs is always deeper than that of the sternal surface, which is often light yellowish-green, with more or less red at the extremities of the forceps. The greenish hue of the sternal surface occasionally passes into yellow in the thorax and into blue in the abdomen.

The distance from the orbit to the posterior margin of the carapace is nearly equal to that from the posterior margin of the carapace to the base of the telson, when the abdomen is fully extended, but this measurement of the carapace is commonly greater than that of the abdomen in the males and less in the females.

The general contour of the carapace (fig. 61), without the rostrum, is that of an oval, truncated at the ends: the anterior end being narrower than the posterior. Its surface is evenly arched from side to side. The greatest breadth of the carapace lies midway between the cervical groove and its posterior edge. Its greatest vertical depth is on a level with the transverse portion of the cervical groove.

The length of the rostrum, measured from the orbit

considerably smaller. Doubtless, the proportional rate of increment is much the same, in the two kinds; but in the English crayfish it has not been actually ascertained.

to its extremity, is greater than half the distance from the orbit to the cervical groove. It is trihedral in section, and its free end is slightly curved upwards (fig. 41). It gradually becomes narrower for about three-fourths of its whole length. At this point it has rather less than half the width which it has at its base (fig. 61, A); and its raised, granular and sometimes distinctly serrated margins are produced into two obliquely directed spines, one on each side. Beyond these, the rostrum rapidly narrows to a fine point; and this part of the rostrum is equal in length to the width between the two spines.

The tergal surface of the rostrum is flattened and slightly excavated from side to side, except in its anterior half, where it presents a granular or finely serrated median ridge, which gradually passes into a low elevation in the posterior half, and, as such, may generally be traced on to the cephalic region of the carapace. The inclined sides of the rostrum meet ventrally in a sharp edge, convex from before backwards; the posterior half of this edge gives rise to a small, usually bifurcated, spine, which descends between the eye-stalks (fig. 41). The raised and granulated lateral margins of the rostrum are continued back on to the carapace for a short distance, as two linear ridges (fig. 61, A). Parallel with each of these ridges, and close to it, there is another longitudinal elevation (*a*, *b*), the anterior end of which is raised into a prominent spine (*a*), which is situated immediately behind the orbit, and may, therefore, be termed the *post-orbital spine.* The elevation itself may be distinguished as the *post-orbital ridge.* The flattened surface of this ridge is marked by a longitudinal depression or groove. The

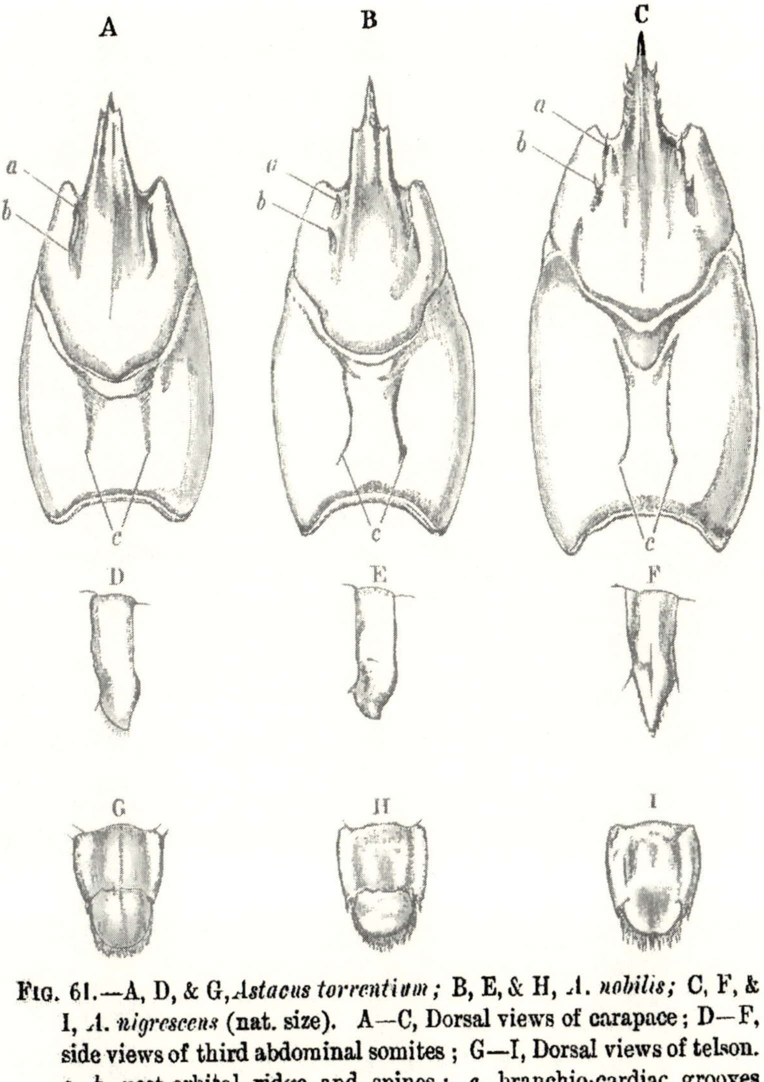

FIG. 61.—A, D, & G, *Astacus torrentium;* B, E, & H, *A. nobilis;* C, F, & I, *A. nigrescens* (nat. size). A—C, Dorsal views of carapace; D—F, side views of third abdominal somites; G—I, Dorsal views of telson. *a*, *b*, post-orbital ridge and spines; *c*, branchio-cardiac grooves inclosing the areola.

posterior end of the ridge passes into a somewhat broader and less marked elevation, the hinder end of which turns inwards, and then comes to an end at a point midway between the orbit and the cervical groove. Generally this hinder elevation appears like a mere continuation of the post-orbital ridge; but, sometimes, the two are separated by a distinct depression. I have never seen any prominent spine upon the posterior elevation, though it is sometimes minutely spinulose. The post-orbital ridges of each side, viewed together, give rise to a characteristic lyrate mark upon the cephalic region of the carapace.

A faintly marked, curved, linear depression runs from the hinder end of the post-orbital ridge, at first directly downwards, and then curves backwards to the cervical groove. It corresponds with the anterior and inferior boundary of the attachment of the adductor muscle of the mandible.

Below the level of this, and immediately behind the cervical groove, there are usually three spines, arranged in a series, which follow the cervical groove. The points of all are directed obliquely forwards, and the lowest is the largest. Sometimes there is only one prominent spine, with one or two very small ones; sometimes there are as many as five of these *cervical spines*.

The cardiac region is marked out by two grooves which run backwards from the cervical groove (fig. 61, A, *c*), and terminate at a considerable distance from the posterior edge of the carapace. Each groove runs, at first, obliquely inwards, and then takes a straight course parallel with its fellow. The area thus defined is termed the *areola*; its breadth is equal to about one-third of the total transverse diameter of the carapace in this region.

No such distinct lines indicate the lateral boundary of the region in front of the cervical groove which answers to the stomach. But the middle part of the carapace, or that which is comprised in the gastric and cardiac regions, has its surface sculptured in a different way from the branchiostegites and the lateral regions of the head. In the former, the surface is excavated by shallow pits, separated by relatively broad flat-topped ridges; but, in the latter, the ridges become more prominent, and take the form of tubercles, the apices of which are directed forwards. Minute setæ spring from the depressions between these tubercles.

The branchiostegite has a thickened rim, which is strongest below and behind (fig. 1). The free edge of this rim is fringed with close-set setæ.

The pleura of the second to the sixth abdominal somites are broadly lanceolate and obtusely pointed at their free ends (fig. 61, D); the anterior edge is longer and more convex than the posterior edge. In the females, the pleura are larger, and are directed more outwards and less downwards than in the males. The pleura of the second somite are much larger than the rest, and overlap the very small pleura of the first somite (fig. 1). The pleura of the sixth somite are narrow, and their posterior edges are concave.

The pits and setæ of the cuticle which clothes the tergal surfaces of the abdominal somites are so few and scattered, that the latter appear almost smooth. In the telson, however, especially in its posterior division, the markings are coarser and the setæ more apparent.

The telson (fig. 61, G) presents an anterior quadrate division and a posterior half-oval part, the free curved edge of which is beset with long setæ, and is sometimes slightly notched in the middle. The posterior division is freely movable upon the anterior, in consequence of the thinness and pliability of the cuticle along a transverse line which joins the postero-external angles of the anterior division, each of which is produced into two strong spines, of which the outer is the longer. The length of the posterior division of the telson, measured from the middle of the suture, is equal to, or but very little less than, that of the anterior division.

On the under side of the head, the basal joints of the antennules are visible, internal to those of the antennæ, but the attachment of the latter is behind and below that of the former (fig. 3, A). Behind these, and in front of the mouth, the epistoma (fig. 39, A, II, III) presents a broad area of a pentagonal form. The posterior boundary of this area is formed by two thickened transverse ridges, which meet on the middle line at a very open angle, the apex of which is turned forwards.

The posterior edges of these ridges are continuous with the labrum. The anterior margin is produced in the middle into a *fleur de lys* shaped process, the summit of which terminates between the antennules. At the sides of this process, the anterior margin of the epistoma is deeply excavated to receive the basal joints of the antennæ. Following the contours of these excavated margins, the surface of the epistoma presents two lateral convexities. The widest and most prominent part of each of these lies towards the outer edge of the epistoma, and is produced into a conical spine. Sometimes there is a second smaller spine beside the principal one. Between the two convexities lies a triangular median depressed area.

The distance from the apex of the anterior median process to the posterior ridge is equal to a little more than half the width of the epistoma.

The corneal surface of the eye is transversely elongated and reniform, and its pigment is black. The eye-stalks are much broader at their bases than at their corneal ends (fig. 48, A). The antennules are about twice as long as the rostrum. The tergal surface of the trihedral basal joint of the antennule, on which the eye-stalk rests, is concave; the outer surface is convex, the inner flat (figs. 26, A, and 48, B). Near the anterior end of the sternal edge which separates the two latter faces, there is a strong curved spine directed forwards (fig. 48, B, *a*). When the setæ, which proceed from the outer edge of the auditory aperture and hide it, are removed, it is seen to be a wide, somewhat triangular cleft, which occupies the greater part of the hinder half of the tergal surface of the basal joint (fig. 26, A).

The exopodites, or squames, of the antennæ extend as far as the apex of the rostrum, or even project beyond it, when they are turned forwards, while they reach to the commencement of the filament of the endopodite (*Frontispiece*). The squame is fully twice as long as it is broad, with a general convexity of its tergal and concavity of its sternal surface. The outer edge is straight and thick, the inner, which is fringed with long setæ, is convex and thin (fig. 48, C). Where these two edges join in front, the squame is produced into a strong spine. A thick outer portion of the squame is marked off from the thinner inner portion by a longitudinal groove on the tergal side, and by a strong ridge on the sternal side. One or two small spines generally project from the posterior and external angle of the squame; but they may be very small or absent in individual specimens. Close beneath these, the outer angle of the next joint is produced into a strong spine. When the abdomen is straightened out, if the antennæ are turned back as far as they will go without damage, the ends of their filaments usually reach the tergum of the third somite of the abdomen (*Frontispiece*). I have not observed any difference between the sexes in this respect.

The inner edge of the ischiopodite of the third maxillipede is strongly serrated and wider in front than behind (fig. 44); the meropodite possesses four or five spines in the same region; and there are one or two spines at the distal end of the carpopodite. When straightened out, the maxillipedes extend as far as, or even beyond, the end of the rostrum.

The inner or sternal edge of the ischiopodite of the forceps is serrated; that of the meropodite presents two rows of spines, the inner small and numerous, the outer large and few. There are several strong spines at the anterior end of the outer or tergal face of this joint. The carpopodite has two strong spines on its under or sternal surface, while its sharp inner edge presents many strong spines. Its upper surface is marked by a longitudinal depression, and is beset with sharp tubercles. The length of the propodite, from its base to the extremity of the fixed claw of the chela, measures rather more than twice as much as the extreme breadth of its base, the thickness of which is less than a third of this length (fig. 20, p. 93). The external angular process, or fixed claw, is of the same length as the base, or a little shorter. Its inner edge is sharp and spinose, and the outer more rounded and simply tuberculated. The apex of the fixed claw is produced into a slightly incurved spine. Its inner edge has a sinuous curvature, convex posteriorly, concave anteriorly, and bears a series of rounded tubercles, of which one near the summit of the convexity, and one near the apex of the claw, are the most prominent.

The apex of the dactylopodite, like that of the propodite, is formed by a slightly incurved spine (fig. 20), while its outer, sharper, edge presents a curvature, the inverse of that of the edge of the fixed claw against which it is applied. This edge is beset with rounded tubercles, the most prominent of which are one at the beginning, and one at the end of the concave posterior moiety of the edge. When the dactylopodite is brought up to the fixed claw, these tubercles lie, one in front of and one behind the chief tubercle of the convexity of the latter. The whole surface of the propodite and dactylopodite is covered with minute elevations, those of the upper surface being much more prominent than those of the lower surface.

The length of the fully extended forceps generally equals the distance between the posterior margin of the orbit and the base of the telson, in well characterized males; and, in individual examples, they are even longer; while it may not be greater than the distance between the orbit and the hinder edge of the fourth abdominal somite, in females; and, in massiveness and strength, the difference of the forceps in the two sexes is still more remarkable (fig. 2). Moreover there is a good deal of variation in the form and size of the chelæ in individual males. The right and left chelæ present no important differences.

The ischiopodites of the four succeeding thoracic limbs are devoid of any recurved spines in either sex (*Front.*, fig. 46). The first pair are the stoutest, the second the longest: and when the latter are spread out at right angles to the body, the distance from tip to tip of the dactylopodites is equal to, or rather greater than, the extreme length of the body from the apex of the rostrum to the posterior edge of the telson, in both sexes. In both sexes, the length of the swimmerets hardly exceeds half the transverse diameter of the somites to which they are attached.

The exopodites of the appendages of the sixth abdominal somite (the extreme length of which is rather greater than that of the telson) are divided into a larger proximal, and a smaller distal portion (fig. 37, F, p. 144). The latter is about half as long as the former, and has a rounded free edge, setose like that of the telson. There is a complete flexible hinge between the two portions, and the overlapping free edge of the proximal portion, which is slightly concave, is beset with conical spines, the outermost of which are the longest. The endopodite has a spine at the junction of its outer straight edge with the terminal setose convex edge. A faintly marked longitudinal median ridge, or keel, ends close to the margin in a minute spine. The tergal distal edge of the protopodite is deeply bilobed, and the inner lobe ends in two spines, while the outer, shorter and broader lobe, is minutely serrated.

In addition to the characters distinctive of sex, which have already been fully detailed (pp. 7, 20, and 145), there is a marked difference in the form of the sterna of the three posterior thoracic somites between the males and females.

Comparing a male and a female of the same size, the triangular area between the bases of the penultimate and ante-penultimate thoracic limbs is considerably broader at the base in the female. In both sexes, the hinder part of the penultimate sternum is a rounded transverse ridge separated by a groove from the anterior part; but this ridge is much larger and more prominent in the female than in the male, and it is often obscurely divided into two lobes by a median depression. Moreover, there are but few setæ on this region in the female; while, in the male, the setæ are long and numerous.

The sternum of the last thoracic somite of the female is divided by a transverse groove into two parts, of which the posterior, viewed from the sternal aspect, has the form of a transverse elongated ridge, which narrows to each end, is moderately convex in the middle, and is almost free from setæ. In the male, the corresponding posterior division of the last thoracic sternum is produced downwards and forwards into a rounded eminence which gives attachment to a sort of brush of long setæ (fig. 35, p. 136).

The importance of this long enumeration of minute details* will appear by and by. It is simply a statement of the more obvious external characters in which all the full-grown English crayfishes which have come under my

* The student of systematic zoology will find the comparison of a lobster with a crayfish in all the points mentioned to be an excellent training of the faculty of observation.

notice agree. No one of these individual crayfishes was exactly like the other; and to give an account of any single crayfish as it existed in nature, its special peculiarities must be added to the list of characters given above; which, considered together with the facts of structure discussed in previous chapters, constitutes a definition, or diagnosis, of the English kind, or *species*, of crayfish. It follows that the species, regarded as the sum of the morphological characters in question and nothing else, does not exist in nature; but that it is an abstraction, obtained by separating the structural characters in which the actual existences—the individual crayfishes—agree, from those in which they differ, and neglecting the latter.

A diagram, embodying the totality of the structural characters thus determined by observation to be common to all our crayfishes, might be constructed; and it would be a picture of nothing which ever existed in nature; though it would serve as a very complete plan of the structure of all the crayfishes which are to be found in this country. The morphological definition of a species is, in fact, nothing but a description of the plan of structure which characterises all the individuals of that species.

California is separated from these islands by a third of the circumference of the globe, one-half of the interval being occupied by the broad North Atlantic ocean. The fresh waters of California, however, contain crayfishes which are so like our own, that it is necessary to compare the two in every point mentioned in the foregoing description in order to estimate the value of the differences which they present. Thus, to take one of the kinds of crayfishes found in California, which has been called *Astacus nigrescens*; the general structure of the animal may be described in precisely the same terms as those used for the English crayfish. Even the branchiæ present no important difference, except that the rudimentary pleurobranchiæ are rather more conspicuous; and that there is a third small one, in front of the two which correspond with those possessed by the English crayfish.

The Californian crayfish is larger and somewhat differently coloured, the undersides of the forceps particularly presenting a red hue. The limbs, and especially the forceps of the males, are relatively longer; the chelæ of the forceps have more slender proportions; the areola is narrower relatively to the transverse diameter of the carapace (fig. 61, C). More definite distinctions are to be found in the rostrum, which is almost parallel-sided for two-thirds of its length, then gives off two strong lateral spines and suddenly narrows to its apex. Behind these spines, the raised lateral edges of the rostrum present five or six other spines which diminish in size from before backwards. The postorbital spine is very prominent, but the ridge is represented, in front, by the base of this spine, which is slightly grooved; and behind, by a distinct spine which is not so strong as the postorbital spine.

There are no cervical spines, and the middle part of the cervical groove is angulated backwards instead of being transverse.

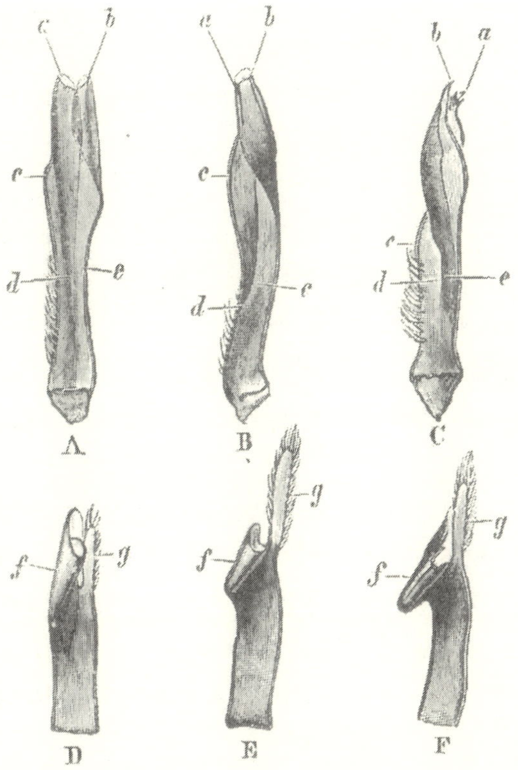

Fig. 62. A & D, *Astacus torrentium;* B & E, *A. nobilis;* C & F, *A. nigrescens.* A—C, 1st abdominal appendage of the male; D—F, endopodite of second appendage (× 3). *a*, anterior, and *b*, posterior rolled edge; *c*, *d*, *e*, corresponding parts of the appendages in each species; *f*, rolled plate of endopodite; *g*, terminal division of endopodite.

The abdominal pleura are narrow, equal-sided, and acutely pointed in the males (fig. 61, F)—slightly broader, more obtuse, and with the anterior edges rather more convex than the posterior, in the females. The tergal surface of the telson is not divided into two parts by a suture (fig. 61, I). The anterior process of the epistoma is of a broad rhomboidal shape, and there are no distinct lateral spines.

The squame of the antenna is not so broad relatively to its length; its inner edge is less convex, and its outer edge is slightly concave; the outer basal angle is sharp but not produced into a spine. The opposed edges of the fixed and movable claws of the chelæ of the forceps are almost straight and present no conspicuous tubercles. In the males, the forceps are vastly larger than in the females, and the two claws of the chelæ are bowed out, so that a wide interval is left when their apices are applied together; in the females, the claws are straight and the edges fit together, leaving no interval. Both the upper and the under surfaces of the claws are almost smooth. The median ridge of the endopodite of the sixth abdominal appendage is more marked, and ends close to the margin in a small prominent spine.

In the females, the posterior division of the sternum of the penultimate thoracic somite is prominent and deeply bilobed; and there are some small differences in form in the abdominal appendages of the males. More especially, the rolled inner process of the endopodite of the second appendage (fig. 62 F, *f*) is disposed very obliquely, and its open mouth is on a level with the base of the jointed part of the endopodite (*g*) instead of reaching almost to the free end of the latter and being nearly parallel with it. In the first appendage (C), the anterior rolled edge (*a*) more closely embraces the posterior (*b*), and the groove is more completely converted into a tube.

It will be observed that the differences between the English and the Californian crayfishes amount to exceedingly little; but, on the assumption that these differences are constant, and that no transitional forms between the English and the Californian crayfishes are to be met with, the individuals which present the characteristic peculiarities of the latter are said to form a distinct species, *Astacus nigrescens;* and the definition of that species is, like that of the English species, a morphological abstraction, embodying an account of the plan of that species, so far as it is distinct from that of other crayfishes.

We shall see by and by that there are sundry other kinds of crayfishes, which differ no more from the English or the Californian kinds, than these do from one another; and, therefore, they are all grouped as species of the one genus, *Astacus.*

If, leaving California, we cross the Rocky Mountains and enter the eastern States of the North American Union, many sorts of crayfishes, which would at once be recognised as such by any English visitor, will be found to be abundant. But on careful examination it will be discovered that all of these differ, both from the English crayfish, and from *Astacus nigrescens*, to a much greater extent than those do from one another. The gills are, in fact, reduced to seventeen on each side, in consequence

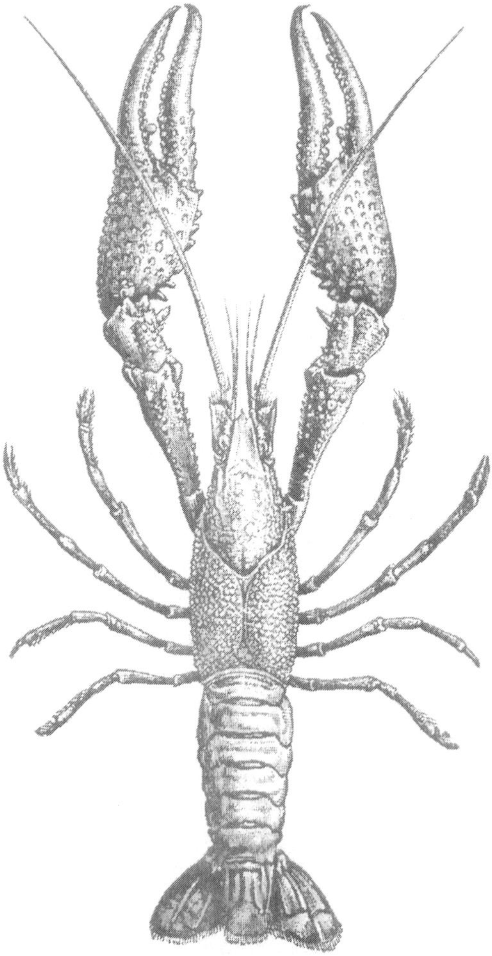

FIG. 63. *Cambarus Clarkii*, male ($\frac{1}{2}$ nat. size), after Hagen.

of the absence of the pleuro-branchia of the last thoracic somite; and there are some other differences to which it is not needful to refer at present. It is convenient to distinguish these seventeen-gilled crayfishes, as a whole, from the eighteen-gilled species; and this is effected by changing the generic name. They are no longer called *Astacus*, but *Cambarus* (fig. 63).

All the individual crayfish referred to thus far, therefore, have been sorted out, first into the groups termed *species;* and then these species have been further sorted into two divisions, termed *genera.* Each genus is an abstraction, formed by summing up the common characters of the species which it includes, just as each species is an abstraction, composed of the common characters of the individuals which belong to it; and the one has no more existence in nature than the other. The definition of the genus is simply a statement of the plan of structure which is common to all the species included under that genus; just as the definition of the species is a statement of the common plan of structure which runs throughout the individuals which compose the species.

Again, crayfishes are found in the fresh waters of the Southern hemisphere; and almost the whole of what has been said respecting the structure of the English crayfish applies to these; in other words, their general plan is the same. But, in these southern crayfishes, the podobranchiæ have no distinct lamina, and the first somite of the abdomen is devoid of appendages in both sexes. The southern crayfishes, like those of the Northern hemisphere, are divisible into many species; and these species are susceptible of being grouped into six genera—*Astacoides* (fig. 65), *Astacopsis*, *Chæraps*, *Parastacus* (fig. 64),

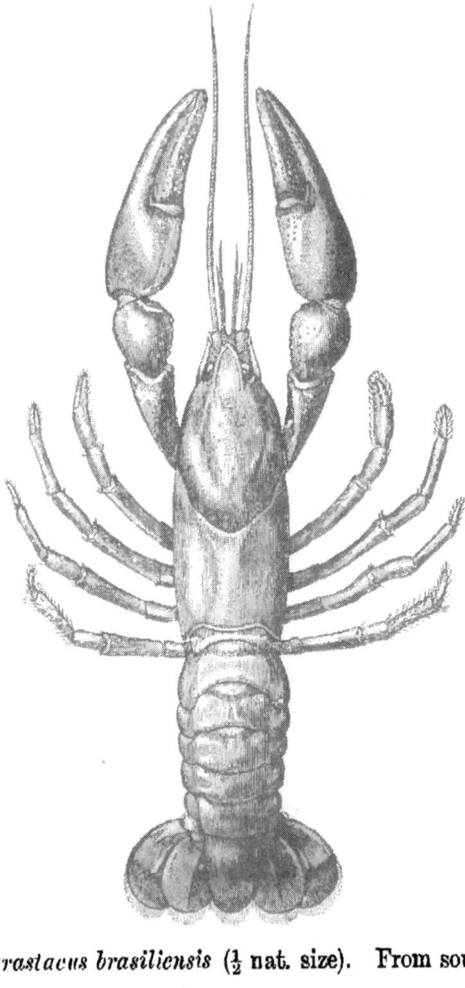

FIG. 64.—*Parastacus brasiliensis* (½ nat. size). From southern Brazil.

Engæus, and *Paranephrops*—on the same principle as that which has led to the grouping of the Northern forms into two genera. But the same convenience which has

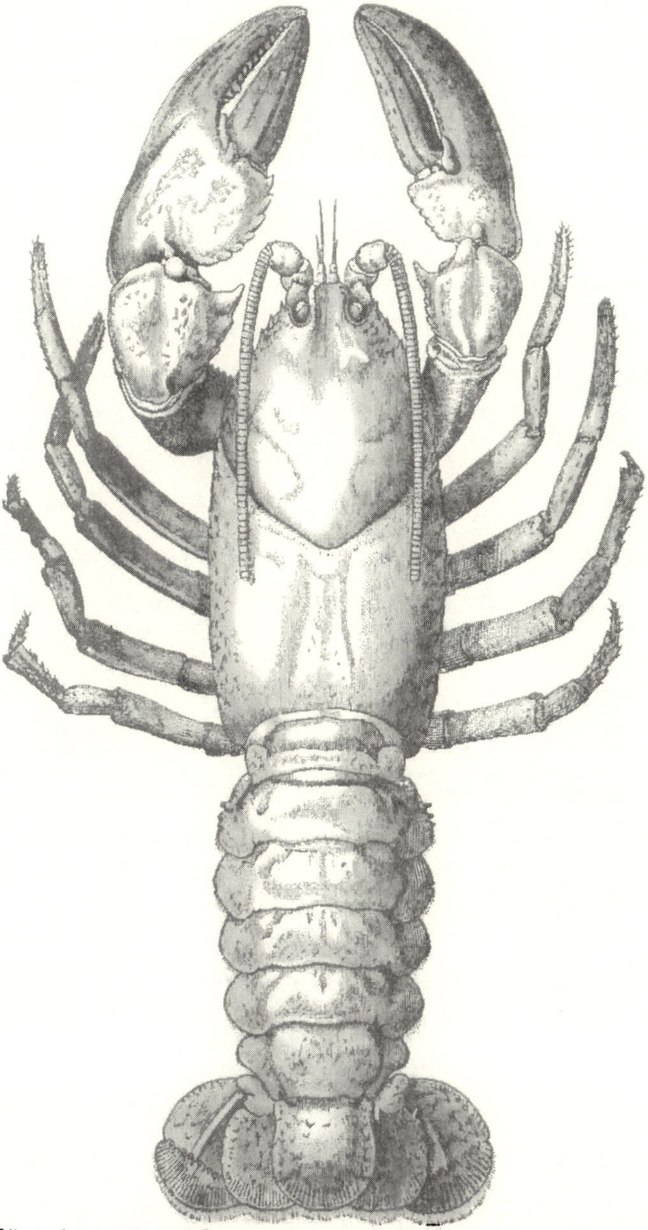

FIG. 65.—*Astacoides madagascarensis* ($\frac{2}{3}$ nat. size). From Madagascar

led to the association of groups of similar species into genera, has given rise to the combination of allied genera into higher groups, which are termed *Families*. It is obvious that the definition of a family, as a statement of the characters in which a certain number of genera agree, is another morphological abstraction, which stands in the same relation to generic, as generic do to specific abstractions. Moreover, the definition of the family is a statement of the plan of all the genera comprised in that family.

The family of the Northern crayfishes is termed *Potamobiidæ;* that of the Southern crayfishes, *Parastacidæ.* But these two families have in common all those structural characters which are special to neither; and, carrying out the metaphorical nomenclature of the zoologist a stage further, we may say that the two form a *Tribe*—the definition of which describes the plan which is common to both families.

It may conduce to intelligibility if these results are put into a graphic form. In fig. 66, A. is a diagram representing the plan of an animal in which all the externally visible parts which are found, more or less modified, in the natural objects which we call individual crayfishes are roughly sketched. It represents the plan of the tribe. B. is a diagram exhibiting such a modification of A. as converts it into the plan common to the whole family of the *Parastacidæ.* C. stands in the same relation to the *Potamobiidæ.* If the scheme were thoroughly worked out, diagrams representing the peculiarities of

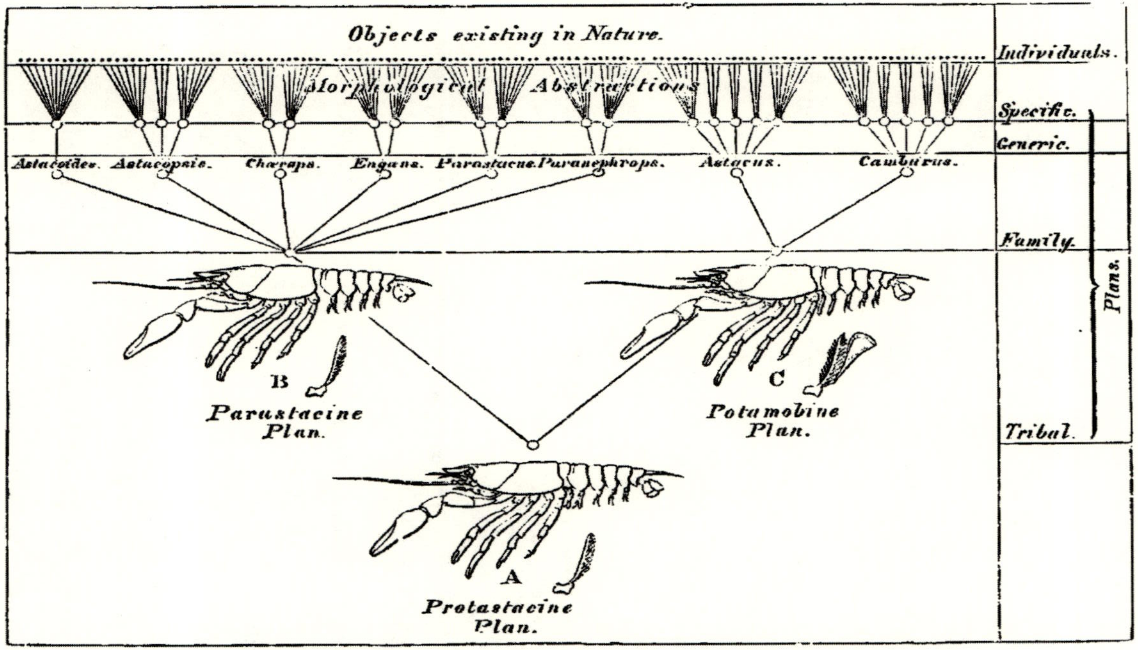

FIG. 66.—Diagram of the morphological relations of the Astacina.

form which characterize each of the genera and species, would appear in the place of the names of the former, or of the circles which represent the latter. All these figures would represent abstractions—mental images which have no existence outside the mind. Actual facts would begin with drawings of individual animals, which we may suppose to occupy the place of the dots above the upper line in the diagram.

That all crayfishes may be regarded as modifications of the common plan A, is not an hypothesis, but a generalization obtained by comparing together the observations made upon the structure of individual crayfishes. It is simply a graphic method of representing the facts which are commonly stated in the form of a definition of the tribe of crayfishes, or *Astacina.*

This definition runs as follows :—

Multicellular animals provided with an alimentary canal and with a chitinous cuticular exoskeleton; with a ganglionated central nervous system traversed by the œsophagus; possessing a heart and branchial respiratory organs.

The body is bilaterally symmetrical, and consists of twenty metameres (or somites and their appendages), of which six are associated into a head, eight into a thorax, and six into an abdomen. A telson is attached to the last abdominal somite.

The somites of the abdominal region are all free, those of the head and thorax, except the hindermost, which is partially free, are united into a cephalothorax, the tergal wall of which has the form of a continuous carapace. The carapace is produced in front into a rostrum, at the sides into branchiostegites.

The eyes are placed at the ends of movable stalks. The antennules are terminated by two filaments. The exopodite of the antenna has the form of a mobile scale. The mandible has a palp. The first and second maxillæ are foliaceous; the second being provided with a large scaphognathite. There are three pairs of maxillipedes, and the endopodites of the third pair are narrow and elongated. The next pair of thoracic appendages is much larger than the rest, and is chelate, as are the two following pairs, which are slender ambulatory limbs. The hindmost two pairs of thoracic appendages are ambulatory limbs, like the foregoing, but not chelate. The abdominal appendages are small swimmerets, except the sixth pair, which are very large, and have the exopodite divided by a transverse joint.

All the crayfishes have a complex gastric armature. The seven anterior thoracic limbs are provided with podobranchiæ, but the first of these is always more or less completely reduced to an epipodite. More or fewer arthrobranchiæ always exist. Pleurobranchiæ may be present or absent.

In this tribe of *Astacina* there are two families, the *Potamobiidæ* and the *Parastacidæ;* and the definition of each of these families is formed by superadding to the definition of the tribe the statement of the special peculiarities of the family.

Thus, the *Potamobiidæ* are those *Astacina* in which the podobranchiæ of the second, fourth, fifth, and sixth thoracic appendages are always provided with a plaited lamina, and that of the first is an epipodite devoid of branchial filaments. The first abdominal somite invariably bears appendages in the males, and usually in both sexes. In the males these appendages are styliform, and those of the second somite are always peculiarly modified. The appendages of the four following somites are relatively small. The telson is very generally divided by a transverse incomplete hinge. None of the branchial filaments are terminated by hooks; nor are any of the coxopoditic setæ, or the longer setæ of the podobranchiæ hooked, though hooked tubercles occur on the stem and on the laminæ of the latter. The coxopoditic setæ are always long and tortuous.

In the *Parastacidæ,* on the other hand, the podobranchiæ are devoid of more than a rudiment of a lamina, though the stem may be alate. The podobranchia of the first maxillipede has the form of an epipodite; but, in almost all cases, it bears a certain number of well developed branchial filaments. The first abdominal somite possesses no appendages in either sex: and the appendages of the four following somites are large. The telson is never divided by a transverse hinge. More or fewer of the branchial filaments of the podobranchiæ are terminated by short hooked spines; and the coxopoditic setæ, as well as those which beset the stems of the podobranchiæ, have hooked apices.

The definitions of the genera would in like manner be given by adding the distinctive characters of each genus to the definitions of the family; and those of the species by adding its character to those of the genus. But at present it is unnecessary to pursue this topic further.

There are no other inhabitants of the fresh waters, or of the land, which could be mistaken for crayfishes; but certain marine animals, familiar to every one, are so strikingly similar to them, that one of these was formerly included in the same genus, *Astacus;* while another is very often known as the "Sea-crayfish." These are the "Common Lobster," the "Norway Lobster," and the "Rock Lobster" or "Spiny Lobster."

The common lobster (*Homarus vulgaris*, fig. 67) presents the following distinctive characters. The last thoracic somite is firmly adherent to the rest; the exopodite of the antenna is so small as to appear like a mere movable scale; all the abdominal appendages are well developed in both sexes; and, in the males, the two anterior pairs are somewhat like those of the male *Astacus*, but less modified.

The principal difference from the *Astacina* is exhibited by the gills, of which there are twenty on each side; namely, six podobranchiæ, ten arthrobranchiæ, and four

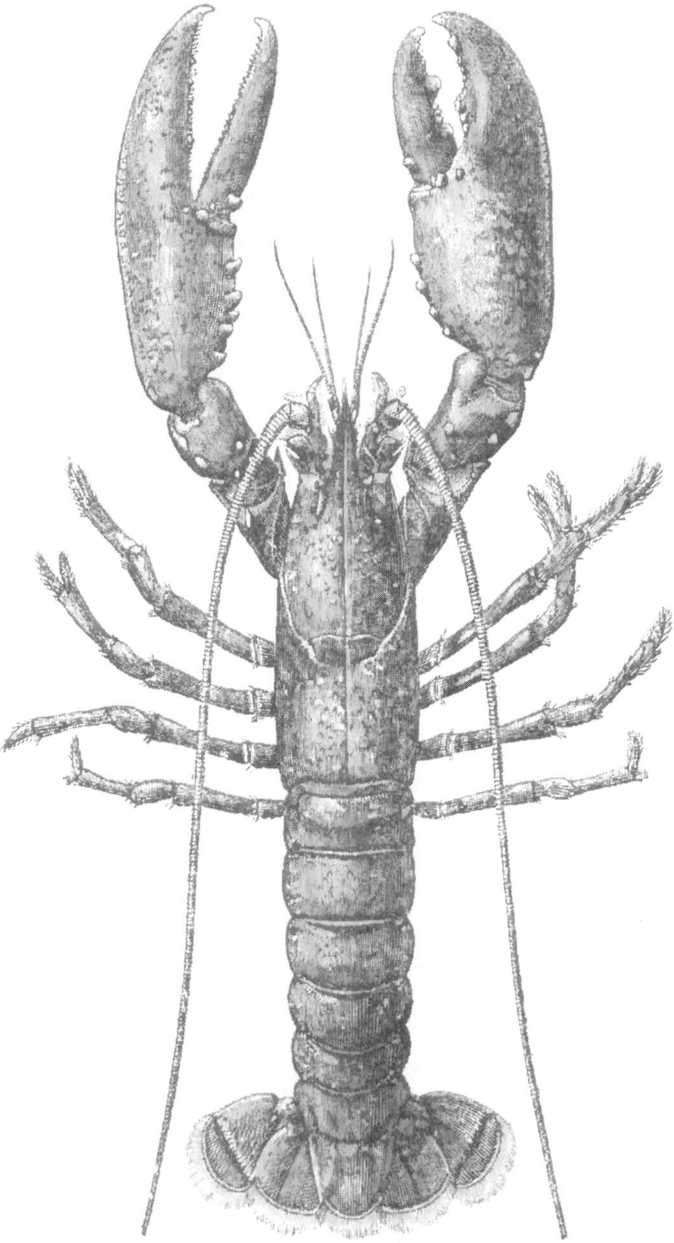

FIG. 67. *Homarus vulgaris* ($\frac{1}{3}$ nat. size).

fully developed pleurobranchiæ. Moreover, the branchial filaments of these gills are much stiffer and more closely set than in most crayfishes. But the most important distinction is presented by the podobranchiæ, in which the stem is, as it were, completely split into two parts longitudinally (as in fig. 68, B); one half (*ep*)

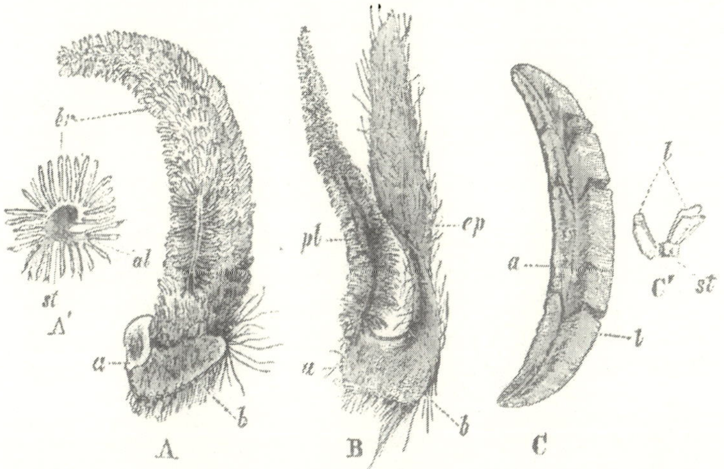

FIG. 68. Podobranchiæ of A, *Parastacus;* B, *Nephrops;* C, *Palæmon.* A', C', transverse sections of A and C respectively. *a*, point of attachment; *al*, wing-like expansion of the stem; *b*, base; *br*, branchial filaments; *ep*, epipodite; *l*, branchial laminæ; *pl*, plume; *st*, stem.

corresponding with the lamina of the crayfish gill, and the other (*pl*) with its plume. Hence the base (*b*) of the podobranchia bears the gill in front; while, behind, it is continued into a broad epipoditic plate (*ep*) slightly folded upon itself longitudinally but not plaited, as in the crayfish.

The Norway Lobster (*Nephrops norvegicus*, fig. 69)

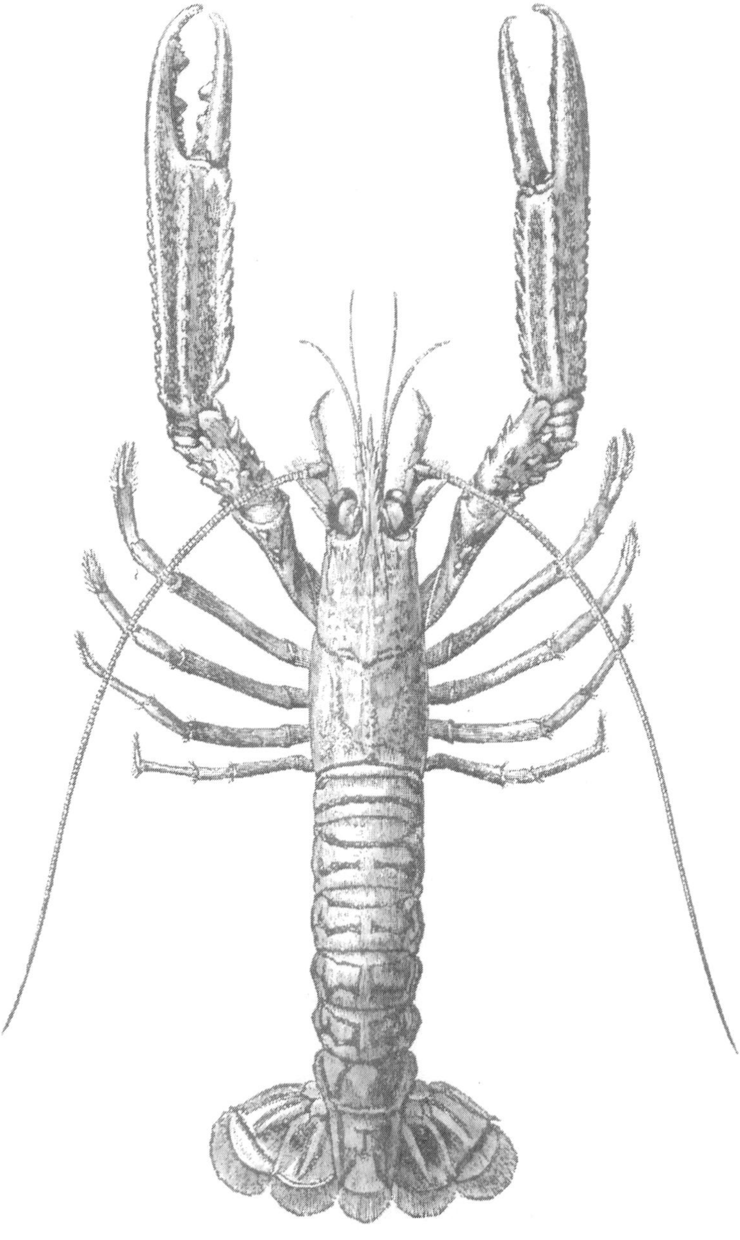

FIG. 69. *Nephrops norvegicus* (½ nat. size).

resembles the lobster in those respects in which the latter differs from the crayfishes: but the antennary squame is large; and, in addition, the branchial plume of the podobranchia of the second maxillipede is very small or absent, so that the total number of functional branchiæ is reduced to nineteen on each side.

These two genera, *Homarus* and *Nephrops*, therefore, represent a family, *Homarina*, constructed upon the same common plan as the crayfishes, but differing so far from the *Astacina* in the structure of the branchiæ and in some other points, that the distinction must be expressed by putting them into a different tribe. It is obvious that the special characteristics of the plan of the *Homarina* give it much more likeness to that of the *Potamobiidæ* than to that of the *Parastacidæ*.

The Rock Lobster (*Palinurus*, fig. 70) differs much more from the crayfishes than either the common lobster or the Norway lobster does. Thus, to refer only to the more important distinctions, the antennæ are enormous; none of the five posterior pairs of thoracic limbs are chelate, and the first pair are not so large in proportion to the rest as in the crayfishes and lobsters. The posterior thoracic sterna are very broad, not comparatively narrow, as in the foregoing genera. There are no appendages to the first somite of the abdomen in either sex. In this respect, it is curious to observe that, in contradistinction from the *Homarina*, the Rock Lobsters are more closely allied to the *Parastacidæ* than to the *Potamobiidæ*.

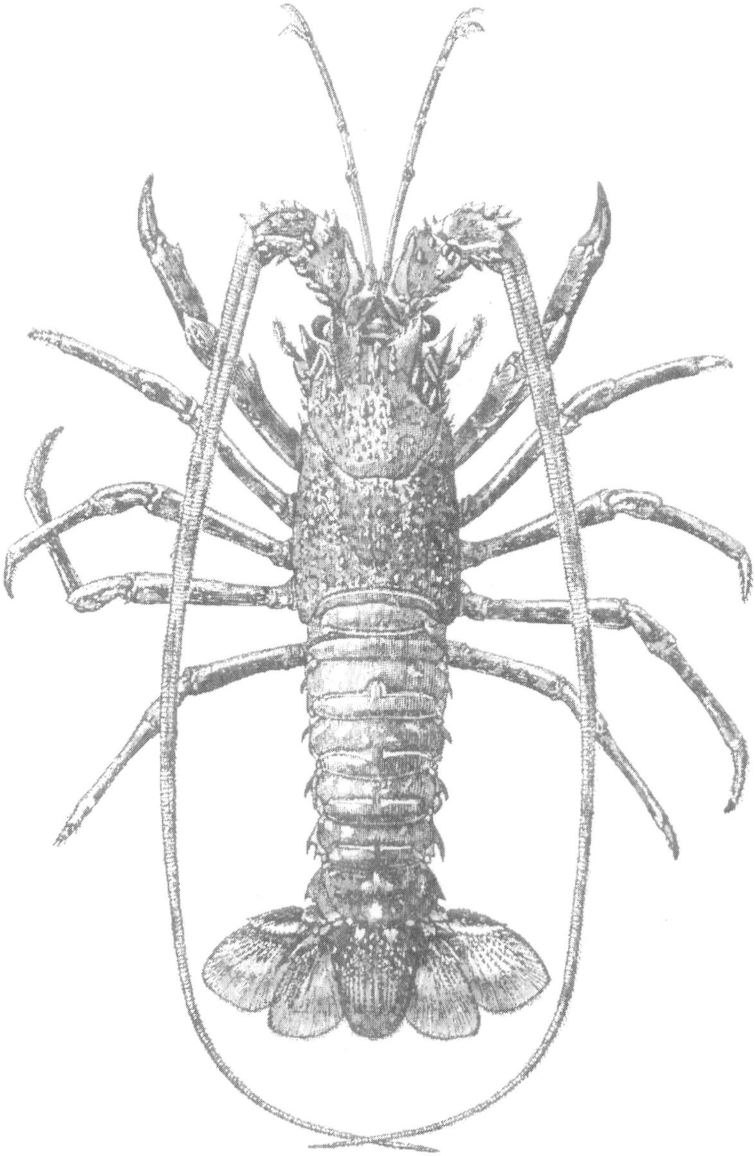

FIG. 70. *Palinurus vulgaris* (about ¼ nat. size).

The gills are similar to those of the lobsters, but reach the number of twenty-one on each side.

In their fundamental structure the rock lobsters agree with the crayfishes; hence the plans of the two may be regarded as modifications of a plan common to both. To this end, the only considerable changes needful in the tribal plan of the crayfishes, are the substitution of simple for chelate terminations to the middle thoracic limbs and the suppression of the appendages of the first somite of the abdomen.

Thus not only all the crayfishes, but all the lobsters and rock lobsters, different as they are in appearance, size, and habits of life, reveal to the morphologist unmistakable signs of a fundamental unity of organization; each is a comparatively simple variation of the general theme—the common plan.

Even the branchiæ, which vary so much in number in different members of these groups, are constructed upon a uniform principle, and the differences which they present are readily intelligible as the result of various modifications of one and the same primitive arrangement.

In all, the gills are *trichobranchiæ;* that is, each gill is somewhat like a bottle-brush, and presents a stem beset, more or less closely, with many series of branchial filaments. The largest number of complete branchiæ possessed by any of the *Potamobiidæ, Parastacidæ, Homaridæ,* or *Palinuridæ,* is twenty-one on each side;

and when this number is present, the total is made up of the same numbers of podobranchiæ, arthrobranchiæ, and pleurobranchiæ attached to corresponding somites. In *Palinurus* and in the genus *Astacopsis* (which is one of the *Parastacidæ*), for example, there are six podobranchiæ attached to the thoracic limbs from the second to the seventh inclusively; five pairs of arthrobranchiæ are attached to the interarticular membranes of the thoracic limbs from the third to the seventh inclusively, and one to that of the second, making eleven in all; while four pleurobranchiæ are fixed to the epimera of the four hindmost thoracic somites. Moreover, in *Astacopsis*, the epipodite of the first thoracic appendage (the first maxillipede) bears branchial filaments, and is a sort of reduced gill.

These facts may be stated in a tabular form as follows:—

The branchial formula of Astacopsis.

Somites and their Appendages.	Podobranchiæ.	Arthrobranchiæ. Anterior.	Arthrobranchiæ. Posterior.	Pleurobranchiæ.		
VII. ...	0 (ep. r.)	0 ...	0 ...	0	=	0 (ep. r.)
VIII. ...	1 ...	1 ...	0 ...	0	=	2
IX. ...	1 ...	1 ...	1 ...	0	=	3
X. ...	1 ...	1 ...	1 ...	0	=	3
XI. ...	1 ...	1 ...	1 ...	1	=	4
XII. ...	1 ...	1 ...	1 ...	1	=	4
XIII. ...	1 ...	1 ...	1 ...	1	=	4
XIV. ...	0 ...	0 ...	0 ...	1	=	1
	—	—	—	—		—
	6 + ep. r. +	6 +	5 +	4	=	21 + ep. r.

This tabular "branchial formula" exhibits at a glance not only the total number of branchiæ, but that of each kind of branchia; and that of all kinds connected with each somite; and it further indicates that the podobranchia of the first thoracic somite has become so far modified, that it is represented only by an epipodite, with branchial filaments scattered upon its surface.

In *Palinurus*, these branchial filaments are absent and the branchial formula therefore becomes—

Somites and their Appendages.	Podobranchiæ.		Arthrobranchiæ. Anterior.		Arthrobranchiæ. Posterior.		Pleurobranchiæ.		
VII. ...	0 (ep.)		0	...	0	...	0	=	0 (ep.)
VIII. ...	1	...	1	...	0	...	0	=	2
IX. ...	1	...	1	...	1	...	0	=	3
X. ...	1	...	1	...	1	...	0	=	3
XI. ...	1	...	1	...	1	...	1	=	4
XII. ...	1	...	1	...	1	...	1	=	4
XIII. ...	1	...	1	...	1	...	1	=	4
XIV. ...	0	...	0	...	0	...	1	=	1
	—		—		—		—		—
	6 + ep.	+	6	+	5	+	4	=	21 + ep.

In the lobster, the solitary arthrobranchia of the eighth somite disappears, and the branchiæ are reduced to twenty on each side.

In *Astacus*, this branchia remains; but, in the English crayfish, the most anterior of the pleurobranchiæ has vanished, and mere rudiments of the two next remain. It has been mentioned that other *Astaci* present a rudiment of the first pleurobranchia.

The branchial formula of Astacus.

Somites and their Appendages.	Podobranchiæ.	Arthrobranchiæ. Anterior.	Arthrobranchiæ. Posterior.	Pleurobranchiæ.	
VII. ...	0 (ep.) ...	0	... 0 ...	0	= 0 (ep.)
VIII. ...	1	... 1	... 0 ...	0	= 2
IX. ...	1	... 1	... 1 ...	0	= 3
X. ...	1	... 1	... 1 ...	0	= 3
XI. ...	1	... 1	... 1 ...	0 or *r*	= 3 or 3 + *r*
XII. ...	1	... 1	... 1 ...	*r*	= 3 + *r*
XIII. ...	1	... 1	... 1 ...	*r*	= 3 + *r*
XIV. ...	0	... 0	... 0 ...	1	= 1
	6 + ep.	+ 6	+ 5 +	1 + 2 or 3 *r*	= 18 + ep. + 2 or 3 *r*.

In *Cambarus*, the number of the branchiæ is reduced to seventeen by the disappearance of the last pleurobranchia; while, in *Astacoides*, the process of reduction is carried so far, that only twelve complete branchiæ are left, the rest being either represented by mere rudiments, or disappearing altogether.

The branchial formula of Astacoides.

Somites and their Appendages.	Podobranchiæ.	Arthrobranchiæ. Anterior.	Arthrobranchiæ. Posterior.	Pleurobranchiæ.	
VII. ...	0 (ep. r.)	0 ...	0 ...	0	= 0 (ep. r.)
VIII. ...	1 ...	*r* ...	0 ...	0	= 1 + *r*
IX. ...	1 ...	1 ...	0 ...	0	= 2
X. ...	1 ...	1 ...	*r* ...	0	= 2 + *r*
XI. ...	1 ...	1 ...	*r* ...	0	= 2 + *r*
XII. ...	1 ...	1 ...	*r* ...	0	= 2 + *r*
XIII. ...	1 ...	1 ...	*r* ...	0	= 2 + *r*
XIV. ...	0 ...	0 ...	0 ...	1	= 1
	6 + *ep. r*	5 + *r* +	0 + 4 *r* +	1	= 12 + ep. r. + 5 *r*.

As these formulæ show, those trichobranchiate crustacea, which possess fewer than twenty-one complete branchiæ on each side, commonly present traces of the missing ones, either in the shape of epipodites, as in the case of the podobranchiæ, or of minute rudiments, in the case of the arthrobranchiæ and the pleurobranchiæ.

In the marine, prawn-like, genus *Penæus* (fig. 73, Chap. VI.), the gills are curiously modified trichobranchiæ. The number of functional branchiæ is, as in the lobster, twenty; but the study of their disposition shows that the total is made up in a very different way.

The branchial formula of Penæus.

Somites and their Appendages.	Podobranchiæ.	Arthrobranchiæ. Anterior.	Arthrobranchiæ. Posterior.	Pleurobranchiæ.	
VII.	0 (ep.)	1	0	0	= 1 + ep.
VIII.	0 (ep.)	1	1	1	= 3 + ep.
IX.	0 (ep.)	1	1	1	= 3 + ep.
X.	0 (ep.)	1	1	1	= 3 + ep.
XI.	0 (ep.)	1	1	1	= 3 + ep.
XII.	0 (ep.)	1	1	1	= 3 + ep.
XIII.	0	1	1	1	= 3
XIV.	0	0	0	1	= 1
	0 + 6 ep. +	7 +	6 +	7	= 20 + 6 ep.

This case is very interesting; for it shows that the whole of the podobranchiæ may lose their branchial character, and be reduced to epipodites, as is the case with the first in the crayfish and lobster, and indeed in most of the forms under consideration. And since all but one of the somites bear both arthrobranchiæ and pleurobranchiæ, the suggestion arises that each hypothetically complete thoracic somite should possess four gills on each side, giving the following

Hypothetically complete branchial formula.

Somites and their Appendages.	Podo-branchiæ.		Arthrobranchiæ. Anterior.		Arthrobranchiæ. Posterior.		Pleuro-branchiæ.		
VII. ...	1	...	1	...	1	...	1	=	4
VIII. ...	1	...	1	...	1	...	1	=	4
IX. ...	1	...	1	...	1	...	1	=	4
X. ...	1	...	1	...	1	...	1	=	4
XI. ...	1	...	1	...	1	...	1	=	4
XII. ...	1	...	1	...	1	...	1	=	4
XIII. ...	1	...	1	...	1	...	1	=	4
XIV. ...	1	...	1	...	1	...	1	=	4
	8	+	8	+	8	+	8	=	32

Starting from this hypothetically complete branchial formula, we may regard all the actual formulæ as produced from it by the more or less complete suppression of the most anterior, or of the most posterior branchiæ, or of both, in each series. In the case of the podobranchiæ, the branchiæ are converted into epipodites; in that of the other branchiæ, they become rudimentary, or disappear.

In general appearance a common prawn (*Palæmon*, fig. 71) is very similar to a miniature lobster or crayfish. Nor does a closer examination fail to reveal a complete fundamental likeness. The number of the somites, and of the appendages, and their general character and disposition, are in fact the same. But, in the prawn, the abdomen is much larger in proportion to the cephalothorax; the

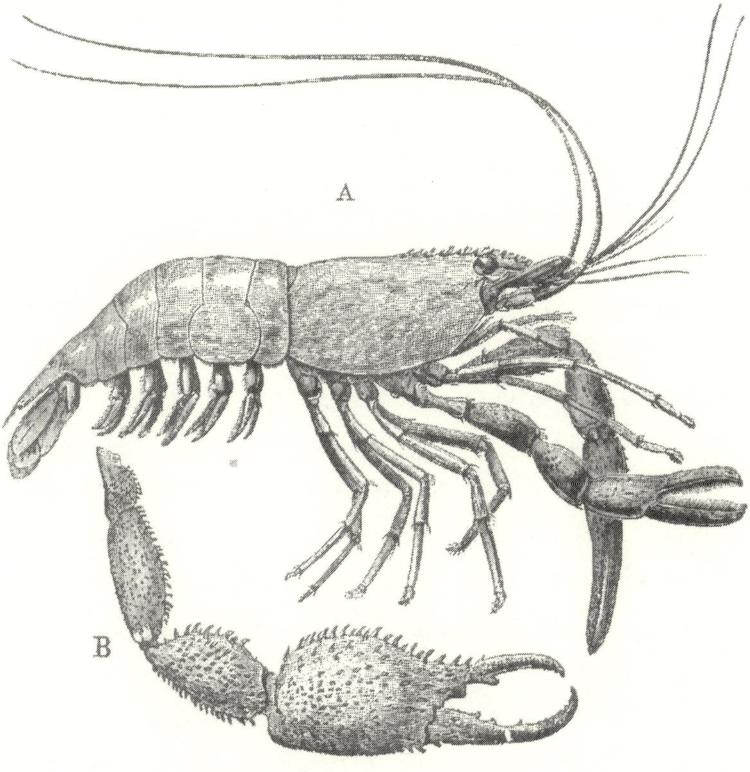

FIG. 71. *Palæmon jamaicensis* (about ⅓ nat. size). A, female; B, fifth thoracic appendage of male.

basal scale, or expodite of the antenna, is much larger; the external maxillipedes are longer, and differ less from the succeeding thoracic appendages. The first pair of these, which answers to the forceps of the crayfish, is chelate, but it is very slender; the second pair, also chelate, is always larger than the first, and is sometimes exceedingly long and strong (fig. 71, B); the remaining thoracic limbs are terminated by simple claws. The five anterior abdominal somites are all provided with large swimmerets, which are used like paddles, when the animal swims quietly; and, in the males, the first pair is only slightly different from the rest. The rostrum is very large, and strongly serrated.

None of these differences from the crayfish, however, is so great, as to prepare us for the remarkable change observable in the respiratory organs. The total number of the gills is only eight. Of these, five are large pleurobranchiæ, attached to the epimera of the five hinder thoracic somites; two are arthrobranchiæ, fixed to the interarticular membrane of the external maxillipede; and one, which is the only complete podobranchia, belongs to the second maxillipede. The podobranchiæ of the first and third maxillipedes are represented only by small epipodites. The branchial formula therefore is:—

Somites and their Appendages.	Podobranchiæ.	Arthrobranchiæ.		Pleurobranchiæ.	
		Anterior.	Posterior.		
VII.	0 (ep.)	0	0	0	= 0 (ep.)
VIII.	1	0	0	0	= 1
IX.	0 (ep.)	1	1	0	= 2 (ep.)
X.	0	0	0	1	= 1
XI.	0	0	0	1	= 1
XII.	0	0	0	1	= 1
XIII.	0	0	0	1	= 1
XIV.	0	0	0	1	= 1
	1 + 2 cp. +	1 +	1 +	5	= 8 + 2 ep.

The prawn, in fact, presents us with an extreme case of that kind of modification of the branchial system, of which *Penæus* has furnished a less complete example. The series of the podobranchiæ is reduced almost to nothing, while the large pleurobranchiæ are the chief organs of respiration.

But this is not the only difference. The prawn's gills are not brush-like, but are foliaceous. They are not *trichobranchiæ*, but *phyllobranchiæ;* that is to say, the central stem of the branchia, instead of being beset with numerous series of slender filaments, bears only two rows of broad flat lamellæ (fig. 68, C, C′, *l*), which are attached to opposite sides of the stem (C′, *s*), and gradually diminish in size from the region of the stem by which it is fixed, upwards and downwards. These lamellæ are superimposed closely upon one another, like the leaves of a book; and the blood traversing the numerous passages by which their substance is excavated, comes into close relation with the currents of aerated water, which are driven between the branchial leaflets by a respiratory mechanism of the same nature as that of the crayfish.

Different as these phyllobranchiæ of the prawns are in appearance from the trichobranchiæ of the preceding *Crustacea,* they are easily reduced to the same type. For in the genus *Axius,* which is closely allied to the lobsters, each branchial stem bears a single series of filaments on its opposite sides; and if these biserial filaments are supposed to widen out into broad leaflets, the transition from the trichobranchia to the phyllobranchia will be very easily effected.

The shrimp (*Crangon*) also possesses phyllobranchiæ, and differs from the prawn chiefly in the character of its locomotive and prehensile thoracic limbs.

There are yet other very well-known marine animals, which, in common appreciation, are always associated with the lobsters and crayfishes, although the difference of general appearance is vastly greater than in any of the cases which have yet been considered. These are the Crabs.

In all the forms we have hitherto been considering, the abdomen is as long as, or longer than, the cephalothorax, while its width is the same, or but little less. The sixth somite has very large appendages, which, together with the telson, make up a powerful tail-fin; and the large abdomen is thus fitted for playing an important part in locomotion.

Again, the length of the cephalothorax is much greater than its width, and it is produced in front into a long rostrum. The bases of the antennæ are freely movable, and they are provided with a movable exopodite. Moreover, the eye-stalks are not inclosed in a cavity or orbit, and the eyes themselves appear above and in front of the antennules. The external maxillipedes are narrow, and their endopodites are more or less leg-like.

None of these statements apply to the crabs. In these

animals the abdomen is short, flattened, and apt to escape immediate notice, as it is habitually kept closely applied against the under surface of the cephalothorax. It is

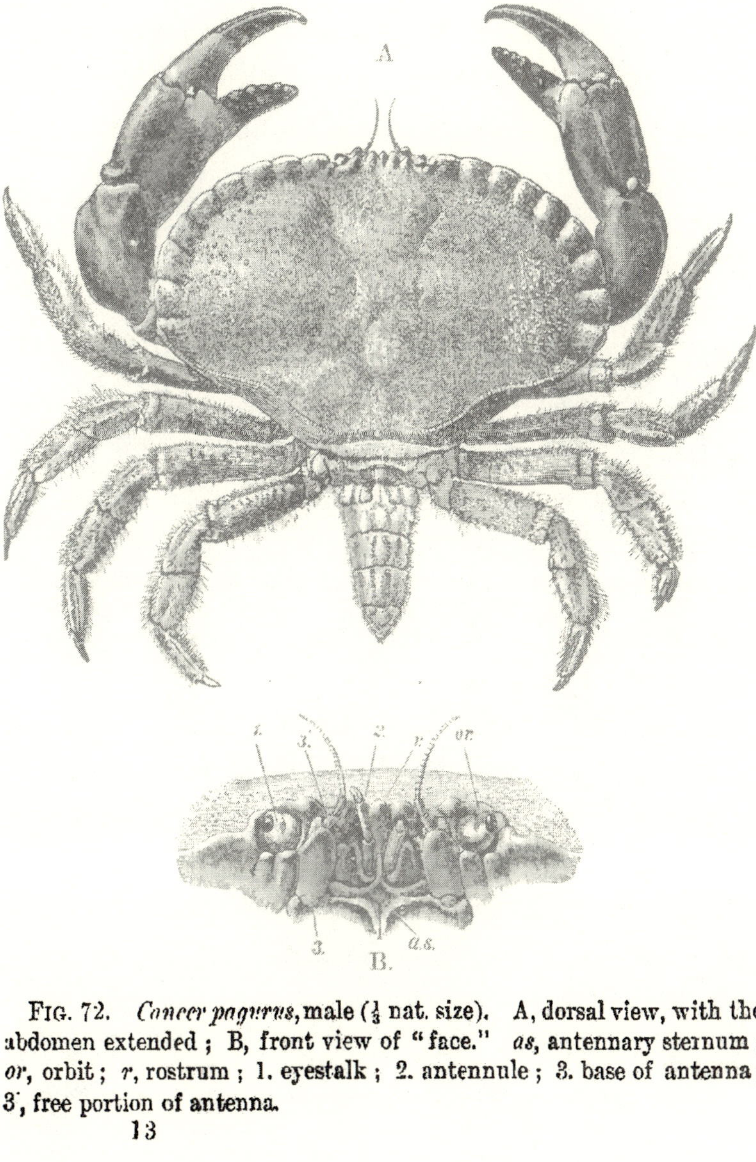

FIG. 72. *Cancer pagurus*, male ($\frac{1}{3}$ nat. size). A, dorsal view, with the abdomen extended; B, front view of "face." *as*, antennary sternum; *or*, orbit; *r*, rostrum; 1. eyestalk; 2. antennule; 3. base of antenna; 3', free portion of antenna.

not used as a swimming organ; and the sixth somite possesses no appendages whatever. The breadth of the cephalothorax is often greater than its length, and there is no prominent rostrum. In its place there is a truncated process (fig. 72, B, *r*), which sends down a vertical partition, and divides from one another two cavities, in which the swollen basal joints of the small antennules (*2*) are lodged. The outer boundary of each of these cavities is formed by the basal part of the antenna (*3*), which is firmly fixed to the edge of the carapace. There is no exopoditic scale; and the free part of the antenna (*3′*) is very small. The convex corneal surface of the eye appears outside the base of the antenna, lodged in a sort of orbit (*or*), the inner margin of which is formed by the base of the antenna, while the upper and outer boundaries are constituted by the carapace. Thus, while in all the preceding forms, the eye is situated nearest the middle line, and is most forward, while the antennule lies outside and behind it, and the antenna comes next; in the crab, the antennule occupies the innermost place, the antenna comes next, and the eye appears to be external to and behind the other two. But there is no real change in the attachments of the eye-stalks. For if the antennule and the basal joint of the antenna are removed, it will be seen that the base of the eye-stalk is attached, as in the crayfish, close to the middle line, on the inner side, and in front of the antennule. But it is very long and extends outwards, behind the antennule and the antenna;

its corneal surface alone being visible, as it projects into the orbit.

Again, the ischiopodites of the external maxillipedes are expanded into broad quadrate plates, which meet in the middle line, and close over the other manducatory organs, like two folding-doors set in a square doorway. Behind these there are great chelate forceps, as in the crayfish; but the succeeding four pairs of ambulatory limbs are terminated by simple claws.

When the abdomen is forcibly turned back, its sternal surface is seen to be soft and membranous. There are no swimmerets; but, in the female, the four anterior pairs of abdominal limbs are represented by singular appendages, which give attachment to the eggs; while in the males there are two pairs of styliform organs attached to the first and second somites of the abdomen, which correspond with those of the male crayfishes.

The ventral portions of the branchiostegites are sharply bent inwards, and their edges are so closely applied throughout the greater part of their length to the bases of the ambulatory limbs, that no branchial cleft is left. In front of the bases of the forceps, however, there is an elongated aperture, which can be shut or opened by a sort of valve, connected with the external maxillipede, which serves for the entrance of water into the branchial cavity. The water employed in respiration, and kept in constant motion by the action of the scaphognathite, is baled out through two apertures, which are separated from the foregoing by the external maxillipedes, and lie at the sides of the quadrate space in which these organs are set.

There are only nine gills on each side, and these, as in the prawn and shrimp, are phyllobranchiæ. Seven of the branchiæ are pyramidal in shape, and for the most part of large size. When the branchiostegite is removed, they are seen lying close against its inner walls, their apices converging towards its summit. The two hindermost of these gills are pleurobranchiæ, the other five are arthrobranchiæ. The two remaining gills are podobranchiæ, and belong to the second and the third maxillipedes respectively. Each is divided into a branchial and an epipoditic portion, the latter having the form of a long curved blade. The branchial portion of the podobranchia of the second maxillipede is long, and lies horizontally under the bases of the four anterior arthrobranchiæ; while the gill of the podobranchia of the third maxillipede is short and triangular, and fits in between the bases of the second and the third arthrobranchiæ. The epipodite of the third maxillipede is very long, and its base furnishes the valve of the afferent aperture of the branchial cavity, which has been mentioned above. The podobranchia of the first maxillipede is represented only by a long curved epipoditic blade, which can sweep over the outer surface of the gills, and doubtless serves to keep them clear of foreign bodies.

The branchial formula of Cancer pagurus.

Somites and their Appendages.		Podobranchiæ.		Arthrobranchiæ. Anterior.		Arthrobranchiæ. Posterior.		Pleurobranchiæ.		
VII.	...	0 (ep.)		0	...	0	...	0	=	0
VIII.	...	1	...	1	...	0	...	0	=	2
IX.	...	1	...	1	...	1	...	0	=	3
X.	...	0	...	1	...	1	...	0	=	2
XI.	...	0	...	0	...	0	...	1	=	1
XII.	...	0	...	0	...	0	...	1	=	1
XIII.	...	0	...	0	...	0	...	0	=	0
XIV.	...	0	...	0	...	0	...	0	=	0
		2 + ep.	+	3	+	2	+	2	=	9 + ep.

It will be observed that the suppression of branchiæ has here taken place in all the series, and at both the anterior and the posterior ends of each. But the defect in total number is made up by the increase of size, not of the pleurobranchiæ alone, as in the case of the prawns, but of the arthrobranchiæ as well. At the same time the whole apparatus has become more specialized and perfected as a breathing organ. The close fitting of the edges of the carapace, and the possibility of closing the inhalent and exhalent apertures, render the crabs much more independent of actual immersion in water than most of their congeners; and some of them habitually live on dry land and breathe by means of the atmospheric air which they take into and expel from their branchial cavities.

Notwithstanding all these wide departures from the structure and habits of the crayfishes, however, attentive examination shows that the plan of construction of the crab is, in all fundamental respects, the same as that of the crayfish. The body is made up of the same number of somites. The appendages of the head and of the thorax are identical in number, in function, and even in the general pattern of their structure. But two pairs of abdominal appendages in the female, and four pairs in the male, have disappeared. The exopodites of the antennæ have vanished, and not even epipodites remain to represent the podobranchiæ of the posterior five pairs of thoracic limbs. The exceedingly elongated eye-stalks are turned backwards and outwards, above the bases of the antennules and the antennæ, and the bases of the latter have become united with the edges of the carapace in front of them. In this manner the extraordinary face, or *metope* (fig. 72, B) of the crab results from a simple modification of the arrangement of parts, every one of which exists in the crayfish. The same common plan serves for both.

The foregoing illustrations are taken from a few of our commonest and most easily obtainable *Crustacea*; but they amply suffice to exemplify the manner in which the conception of a plan of organization, common to a multitude of animals of extremely diverse outward forms and habits, is forced upon us by mere comparative anatomy.

Nothing would be easier, were the occasion fitting, than to extend this method of comparison to the whole of the several thousand species of crab-like, crayfish-like, or prawn-like animals, which, from the fact that they all have their eyes set upon movable stalks, are termed the *Podophthalmia*, or stalk-eyed *Crustacea*; and by arguments of similar force to prove that they are all modifications of the same common plan. Not only so, but the sand-hoppers of the sea-shore, the wood-lice of the land, and the water-fleas or the monoculi of the ponds, nay, even such remote forms as the barnacles which adhere to floating wood, and the acorn shells which crowd every inch of rock on many of our coasts, reveal the same fundamental organization. Further than this, the spiders and the scorpions, the millipedes and the centipedes, and the multitudinous legions of the insect world, show us, amid infinite diversity of detail, nothing which is new in principle to any one who has mastered the morphology of the crayfish.

Given a body divided into somites, each with a pair of appendages; and given the power to modify those somites and their appendages in strict accordance with the principles by which the common plan of the *Podophthalmia* is modified in the actually existing members of that order; and the whole of the *Arthropoda*, which probably make up two-thirds of the animal world, might readily be educed from one primitive form.

And this conclusion is not merely speculative. As a matter of observation, though the *Arthropoda* are not all evolved from one primitive form, in one sense of the words, yet they are in another. For each can be traced back in the course of its development to an ovum, and that ovum gives rise to a blastoderm, from which the parts of the embryo arise in a manner essentially similar to that in which the young crayfish is developed.

Moreover, in a large proportion of the *Crustacea*, the embryo leaves the egg under the form of a small oval body, termed a *Nauplius* (fig. 73, D), provided with (usually) three pairs of appendages, which play the part of swimming limbs, and with a median eye. Changes of form accompanied by sheddings of the cuticle take place, in virtue of which the larva passes into a new stage, when it is termed a *Zoœa* (C). In this, the three pairs of locomotive appendages of the *Nauplius* are metamorphosed into rudimentary antennules, antennæ, and mandibles, while two or more pairs of anterior thoracic appendages provided with exopodites and hence appearing bifurcated, subserve locomotion. The abdomen has grown out and become a notable feature of the Zoæa, but it has no appendages.

In some *Podophthalmia*, as in *Penæus* (fig. 73), the young leaves the egg as a Nauplius, and the Nauplius becomes a Zoæa. The hinder thoracic appendages, each provided with an epipodite, appear; the stalked eyes and the abdominal members are developed, and the larva passes into what is sometimes called the *Mysis* or *Schizopod* stage. The adult state differs from this chiefly in the presence of branchiæ and the rudimentary character of the exopodites of the five posterior thoracic limbs.

In the Opossum-shrimps (*Mysis*) the young does not leave the pouch of the mother until it is fully

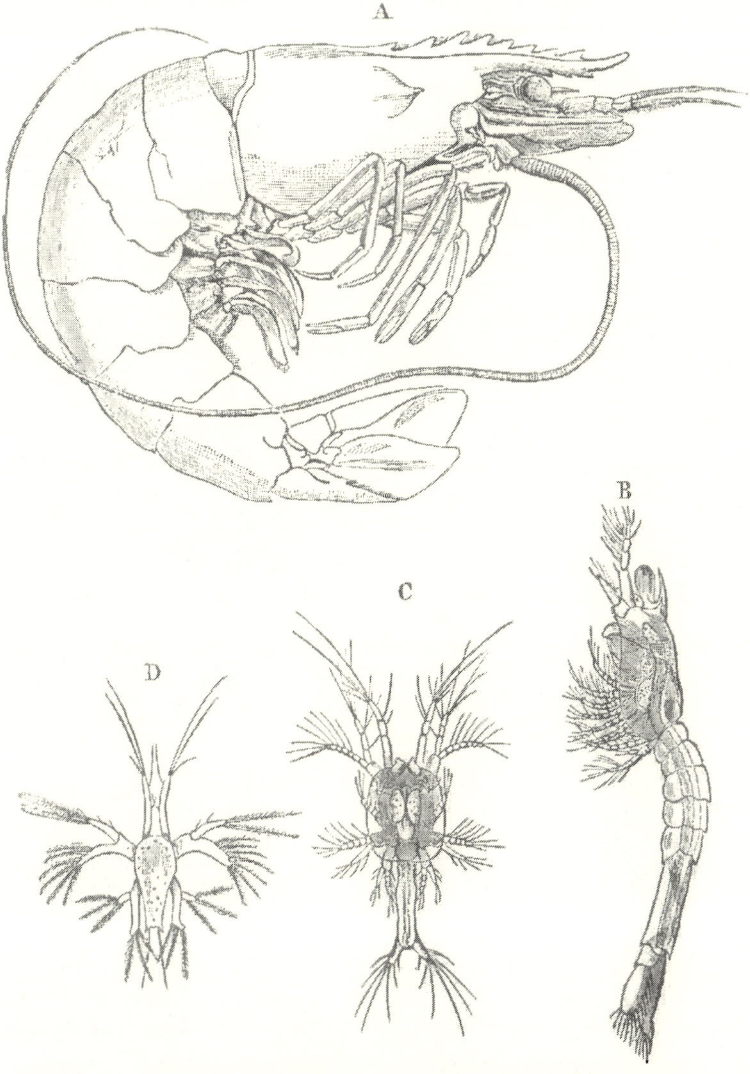

FIG. 73. *Penæus semisulcatus.* A, adult (after de Haan, ½ nat. size); B, Zoæa, and C, less advanced Zoæa of a species of *Penæus.* D Nauplius. (B, C, and D, after Fritz Müller.)

developed; and, in this case, the *Nauplius* state is passed through so rapidly and in so early and imperfect a condition of the embryo, that it would not be recognized

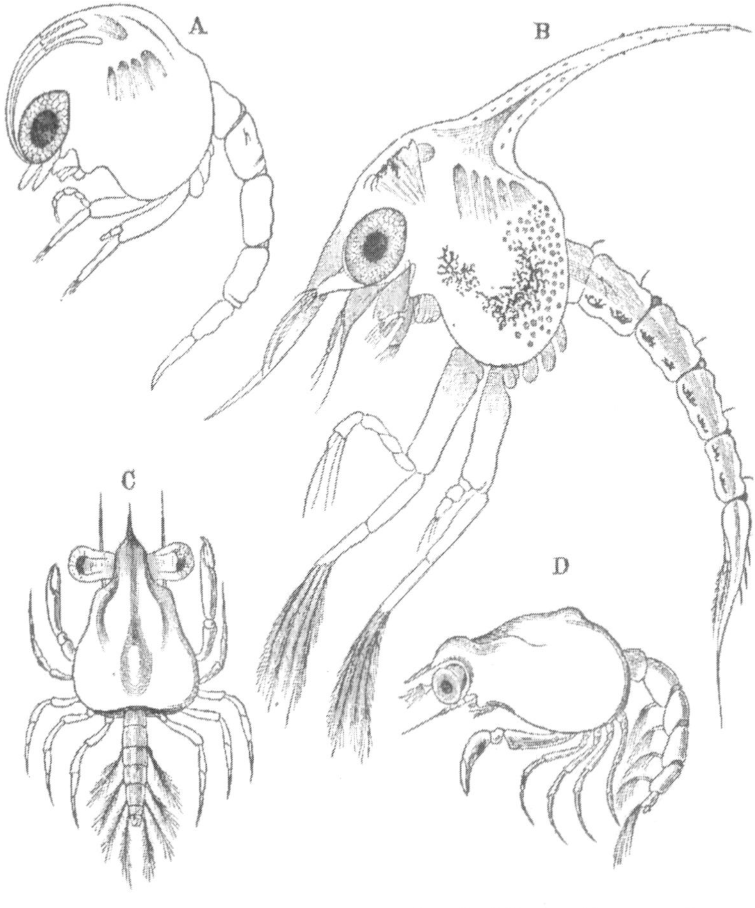

FIG. 74. *Cancer pagurus.* A, newly hatched Zoæa; B, more advanced Zoæa; C, dorsal, and D, side view of Megalopa (after Spence Bate). The figures A and B are more magnified than C and D.)

except for the cuticle which is developed and is subsequently shed.

In the great majority of the *Podophthalmia*, the Nauplius stage seems to be passed over without any such clear evidence of its occurrence, and the young is set free as a Zoæa. In the lobsters, which have, throughout life, a large abdomen provided with swimmerets, the Zoæa, after going through a Mysis or Schizopod stage, passes into the adult form.

In the crab, the young leaves the egg as a Zoæa (fig. 74, A and B). But this is not followed by a Schizopod stage, inasmuch as the five hinder pair of thoracic limbs are apparently, from the first, devoid of exopodites. But the Zoæa, after it has acquired stalked eyes and a complete set of thoracic and abdominal members, and has passed into what is called the *Megalopa* stage (fig. 74, C and D), suffers a more complete metamorphosis. The carapace widens, the fore part of the head is modified so as to bring about the formation of the characteristic metope: and the abdomen, losing more or fewer of its posterior appendages, takes up its final position under the thorax.

In the Zoæa state, those thoracic limbs which give rise to the maxillipedes are provided with well-developed exopodites, and in the free Mysis state all these limbs have exopodites. In the Opossum-shrimps these persist throughout life; in *Penæus*, the rudiments of them only remain; in the lobster, they disappear altogether.

Thus, in these animals, there is no difficulty in demonstrating that embryological uniformity of type of all the limbs, complete evidence of which was not furnished by the development of the crayfish. In this crustacean, in fact, it would appear that the process of development has undergone its maximum of abbreviation. The embryo presents no distinct and independent Nauplius or Zoæa stages, and, as in the crab, there is no Schizopod or Mysis stage. The abdominal appendages are developed very early, and the new born young, which resembles the Megalopa stage of the crab, differs only in a few points from the adult animal.

Guided by comparative morphology, we are thus led to admit that the whole of the *Arthropoda* are connected by closer or more remote degrees of affinity with the crayfish. If we were to study the perch and the pond-snail with similar care, we should be led to analogous conclusions. For the perch is related by similar gradations, in the first place, with other fishes; then more remotely, with frogs and newts, reptiles, birds, and mammals; or, in other words, with the whole of the great division of the *Vertebrata*. The pond-snail, by like reasoning upon analogous data, is connected with the *Mollusca*, in all their innumerable kinds of slugs, shellfish, squids, and cuttlefish. And, in each case, the study of development takes us back to an egg as the primary condition of the animal, and to the process of yelk division, the formation of a blastoderm, and the conversion of that blastoderm into a more or less modified gastrula, as the early stages of development. The like is true of all the worms, sea-urchins, starfishes, jellyfishes, polypes, and sponges; and it is only in the minutest and simplest forms of animal life that the germ, or representative of the ovum becomes metamorphosed into the adult form without the preliminary process of division.

In the majority even of these *Protozoa*, the typical structure of the nucleated cell is retained, and the whole animal is the equivalent of a histological unit of one of the higher organisms. An *Amœba* is strictly comparable, morphologically, to one of the corpuscles of the blood of the crayfish.

Thus, to exactly the same extent as it is legitimate to represent all the crayfishes as modifications of the common astacine plan, it is legitimate to represent all the multicellular animals as modifications of the gastrula, and the gastrula itself as a peculiarly disposed aggregate of cells; while the *Protozoa* are such cells either isolated, or otherwise aggregated.

It is easy to demonstrate that all plants are either cell aggregates, or simple cells; and as it is impossible to draw any precise line of demarcation, either physiological or morphological, between the simplest plants, and the simplest of the *Protozoa*, it follows that all forms of life are morphologically related to one another; and that in whatever sense we say that the English and the Californian crayfish are allied, in the same sense, though not to the same degree, must we admit that all living things are allied. Given one of those protoplasmic bodies, of which we are unable to say certainly whether it is animal or plant, and endow it with such inherent capacities of self-modification as are manifested daily under our eyes by developing ova, and we have a sufficient reason for the existence of any plant, or of any animal.

This is the great result of comparative morphology; and it is carefully to be noted that this result is not a speculation, but a generalisation. The truths of anatomy and of embryology are generalised statements of facts of experience; the question whether an animal is more or less like another in its structure and in its development, or not, is capable of being tested by observation; the doctrine of the unity of organisation of plants and animals is simply a mode of stating the conclusions drawn from experience. But, if it is a just mode of stating these conclusions, then it is undoubtedly conceivable that all plants and all animals may have been evolved from a common physical basis of life, by processes similar to those which we every day see at work in the evolution of individual animals and plants from that foundation.

That which is conceivable, however, is by no means necessarily true; and no amount of purely morphological evidence can suffice to prove that the forms of life have come into existence in one way rather than another.

There is a common plan among churches, no less than

among crayfishes; nevertheless the churches have certainly not been developed from a common ancestor, but have been built separately. Whether the different kinds of crayfishes have been built separately, is a problem we shall not be in a position to grapple with, until we have considered a series of facts connected with them, which have not yet been touched upon.

CHAPTER VI.

THE DISTRIBUTION AND THE ÆTIOLOGY OF THE CRAYFISHES.

So far as I have been able to discover, all the crayfishes which inhabit the British islands agree in every point with the full description given above, at p. 230. They are abundant in some of our rivers, such as the Isis, and other affluents of the Thames; and they have been observed in those of Devon;* but they appear to be absent from many others. I cannot hear of any, for example, in the Cam or the Ouse, on the east, or in the rivers of Lancashire and Cheshire, on the west. It is still more remarkable that, according to the best information I can obtain, they are absent in the Severn, though they are plentiful in the Thames and Severn canal. Dr. M'Intosh, who has paid particular attention to the fauna of Scotland, assures me that crayfish are unknown north of the Tweed. In Ireland, on the other hand, they occur in many localities; † but the question whether their diffusion, and even their introduction into this

* Moore. Magazine of Natural History. New Series, III., 1839.

† Thompson. Annals and Magazine of Natural History, XI., 1843.

island, has or has not been effected by artificial means, is involved in some obscurity.

English zoologists have always termed our crayfish *Astacus fluviatilis;* and, up to a recent period, the majority of Continental naturalists have included a corresponding form of *Astacus* under that specific name.

Thus M. Milne Edwards, in his classical work on the *Crustacea*,* published in 1837, observes under the head of "Écrevisse commune. *Astacus fluviatilis*:" "There are two varieties of this crayfish; in the one, the rostrum gradually becomes narrower from its base onwards, and the lateral spines are situated close to its extremity; in the other, the lateral edges of the rostrum are parallel in their posterior half and the lateral spines are stronger and more remote from the end."

The "first variety," here mentioned, is known under the name of "Écrevisse à pieds blancs"† in France, by way of distinction from the "second variety," which is termed "Écrevisse à pieds rouges," on account of the more or less extensive red coloration of the forceps and ambulatory limbs. This second variety is the larger, commonly attaining five inches in length, and sometimes reaching much larger dimensions; and it is more highly esteemed for the market, on account of its better flavour.

In Germany, the two forms have long been popularly distinguished, the former by the name of "Steinkrebs,"

* "Histoire Naturelle des Crustacés."

† Carbonnier. "L'Écrevisse," p. 8.

or "stone crayfish," and the latter by that of "Edelkrebs," or "noble crayfish."

Milne Edwards, it will be observed, speaks of these two forms of crayfish as "varieties" of the species *Astacus fluviatilis*; but, even as far back as the year 1803 some zoologists began to regard the "stone crayfish" as a distinct species, to which Schrank applied the name of *Astacus torrentium*, while the "noble crayfish" remained in possession of the old denomination, *Astacus fluviatilis*; and, subsequently, various forms of "stone-crayfishes" have been further distinguished as the species *Astacus saxatilis*, *A. tristis*, *A. pallipes*, *A. fontinalis*, &c. On the other hand, Dr. Gerstfeldt,* who has devoted especial attention to the question, denies that these are anything more than varieties of one species; but he holds this and Milne Edwards's "second variety" to be specifically distinct from one another.

We thus find ourselves in the presence of three views respecting the English and French crayfishes.

1. They are all varieties of one species—*A. fluviatilis*.

2. There are two species—*A. fluviatilis*, and *A. torrentium*, of which last there are several varieties.

3. There are, at fewest, five or six distinct species.

Before adopting the one or the other of these views, it is necessary to form a definite conception of the meaning of the terms "species" and "variety."

* "Ueber die Flusskrebse Europas." Mém. de l'Acad. de St. Petersburg, 1859.

The word "species" in Biology has two significations; the one based upon morphological, the other upon physiological considerations.

A species, in the strictly morphological sense, is simply an assemblage of individuals which agree with one another, and differ from the rest of the living world in the sum of their morphological characters; that is to say, in the structure and in the development of both sexes. If the sum of these characters in one group is represented by A, and that in another by A + *n*; the two are morphological species, whether *n* represents an important or an unimportant difference.

The great majority of species described in works on Systematic Zoology are merely morphological species. That is to say, one or more specimens of a kind of animal having been obtained, these specimens have been found to differ from any previously known by the character or characters *n*; and this difference constitutes the definition of the new species, and is all we really know about its distinctness.

But, in practice, the formation of specific groups is more or less qualified by considerations based upon what is known respecting variation. It is a matter of observation that progeny are never exactly like their parents, but present small and inconstant differences from them. Hence, when specific identity is predicated of a group of individuals, the meaning conveyed is not that they are all exactly alike, but only that their differences are so small, and so inconstant, that they lie within the probable limits of individual variation.

Observation further acquaints us with the fact, that, sometimes, an individual member of a species may exhibit a more or less marked variation, which is propagated through all the offspring of that individual, and may even become intensified in them. And, in this manner, a *variety*, or *race*, is generated within the species; which variety, or race, if nothing were known respecting its origin, might have every claim to be regarded as a separate morphological species. The distinctive characters, of a race, however, are rarely equally well marked in all the members of the race. Thus suppose the species A to develope the race A + x; then the difference x is apt to be much less in some individuals than in others; so that, in a large suite of specimens, the interval between A + x and A will be filled up by a series of forms in which x gradually diminishes.

Finally, it is a matter of observation that modification of the physical conditions under which a species lives favours the development of varieties and races.

Hence, in the case of two specimens having respectively the characters A and A + n, although, *primâ facie*, they are of distinct species; yet if a large collection shows us that the interval between A and A + n is filled up by forms of A having traces of n, and forms of A + n in which n becomes less and less, then it will be concluded that A and A + *n* are races of one species and not separate species. And this conclusion will be fortified if A and A + *n* occupy different stations in the same geographical area.

Even when no transitional forms between A and A + *n* are discoverable, if *n* is a small and unimportant difference, such as of average size, colour, or ornamentation, it may be fairly held that A and A + *n* are mere varieties; inasmuch as experience proves that such variations may take place comparatively suddenly; or the intermediate forms may have died out and thus the evidence of variation may have been effaced.

From what has been said it follows that the groups termed morphological species are provisional arrangements, expressive simply of the present state of our knowledge.

We call two groups species, if we know of no transitional forms between them, and if there is no reason to believe that the differences which they present are such as may arise in the ordinary course of variation. But it is impossible to say whether the progress of inquiry into the characters of any group of individuals may prove that what have hitherto been taken for mere varieties are distinct morphological species; or whether, on the contrary, it may prove that what have hitherto been regarded as distinct morphological species are mere varieties.

What has happened in the case of the crayfish is this:

the older observers lumped all the Western European forms which came under their notice under one species, *Astacus fluviatilis;* noting, more or less distinctly, the stone crayfish and the noble crayfish as races or varieties of that species. Later zoologists, comparing crayfishes together more critically, and finding that the stone crayfish is ordinarily markedly different from the noble crayfish, concluded that there were no transitional forms, and made the former into a distinct species, tacitly assuming that the differential characters are not such as could be produced by variation.

It is at present an open question whether further investigation will or will not bear out either of these assumptions. If large series of specimens of both stone crayfishes and noble crayfishes from different localities are carefully examined, they will be found to present great variations in size and colour, in the tuberculation of the carapace and limbs, and in the absolute and relative sizes of the forceps.

The most constant characters of the stone crayfish are:—

1. The tapering form of the rostrum and the approximation of the lateral spines to its point; the distance between these spines being about equal to their distance from the apex of the rostrum (fig. 61, A).

2. The development of one or two spines from the ventral margin of the rostrum.

3. The gradual subsidence of the posterior part of

the post-orbital ridge, and the absence of spines on its surface.

4. The large relative size of the posterior division of the telson (G).

On the contrary, in the noble crayfish :—

1. The sides of the posterior two-thirds of the rostrum are nearly parallel, and the lateral spines are fully a third of the length of the rostrum from its point; the distance between them being much less than their distance from the apex of the rostrum (B).

2. No spine is developed from the ventral margin of the rostrum.

3. The posterior part of the post-orbital ridge is a more or less distinct, sometimes spinous elevation.

4. The posterior division of the telson is smaller relatively to the anterior division (H).

I may add that I have found three rudimentary pleurobranchiæ in the noble crayfish, and never more than two in the stone crayfish.

In order to ascertain whether no crayfish exist in which the characters of the parts here referred to are intermediate between those defined, it would be necessary to examine numerous examples of each kind of crayfish from all parts of the areas which they respectively inhabit. This has been done to some extent, but by no means thoroughly; and I think that all that can be safely said, at present, is that the existence of intermediate forms is not proven. But, whatever the constancy of the differences between the two kinds of crayfishes, there can surely be no doubt as to their insignificance; and no question that they are no more than such as, judging by analogy, might be produced by variation.

From a morphological point of view, then, it is really impossible to decide the question whether the stone crayfish and the noble crayfish should be regarded as species or as varieties. But, since it will, hereafter, be convenient to have distinct names for the two kinds, I shall speak of them as *Astacus torrentium* and *Astacus nobilis.**

In the physiological sense, a species means, firstly, a group of animals the members of which are capable of completely fertile union with one another, but not with the members of any other group; and, secondly, it means all the descendants of a primitive ancestor or ancestors, supposed to have originated otherwise than by ordinary generation.

It is clear that, even if crayfishes had an unbegotten ancestor, there is no means of knowing whether the stone crayfish and the noble crayfish are descendants of the same, or of different ancestors, so that the second sense of species hardly concerns us. As to the first sense, there is no evidence to show whether the two

* According to strict zoological usage the names should be written *A. fluviatilis* (var. *torrentium*) and *A. fluviatilis* (var. *nobilis*) on the hypothesis that the stone crayfish and the noble crayfish are varieties; and *A. torrentium* and *A. fluviatilis* on the hypothesis that they are species; but as I neither wish to prejudge the species question, nor to employ cumbrously long names, I take a third course.

kinds of crayfish under consideration are capable of fertile union or whether they are sterile. It is said, however, that hybrids or mongrels are not met with in the waters which are inhabited by both kinds, and that the breeding season of the stone crayfish begins earlier than that of the noble crayfish.

M. Carbonnier, who practises crayfish culture on a large scale, gives some interesting facts bearing on this question in the work already cited. He says that, in the streams of France, there are two very distinct kinds of crayfishes—the red-clawed crayfish (L'Écrevisse à pieds rouges), and the white-clawed crayfish (L'Écrevisse à pieds blancs), and that the latter inhabit the swifter streams. In a piece of land converted into a crayfish farm, in which the white-clawed crayfish existed naturally in great abundance, 300,000 red-clawed crayfish were introduced in the course of five years; nevertheless, at the end of this time, no intermediate forms were to be seen, and the "pieds rouges" exhibited a marked superiority in size over the "pieds blancs." M. Carbonnier, in fact, says that they were nearly twice as big.

On the whole, the facts as at present known, seem to incline rather in favour of the conclusion that *A. torrentium* and *A. nobilis* are distinct species; in the sense that transitional forms have not been clearly made out, and that, possibly, they do not interbreed.

As I have already remarked, the very numerous

specimens of English and Irish crayfishes which have passed through my hands, have all presented the character of *Astacus torrentium*, with which also the description given in works of recognised authority coincides as far as it goes.* The same form is found in many parts of France, as far south as the Pyrenees, and it is met with as far east as Alsace and Switzerland. I have recently† been enabled, by the kindess of Dr. Bolivar, of Madrid, who sent me a number of crayfishes from the neighbourhood of that city, to satisfy myself that the Spanish peninsula contains crayfishes altogether similar to those of Britain, except that the subrostral spine is less developed. Further, I have no doubt that Dr. Heller‡ is right in his identification of the English crayfish with a form which he describes under the name of *A. saxatilis*. He says that it is especially abundant in Southern Europe, and that it occurs in Greece, in Dalmatia, in the islands of Cherso and Veglia, at Trieste, in the Lago di Garda, and at Genoa. Further, *Astacus torrentium* appears to be widely distributed in North Germany. The eastern limit of this crayfish is uncertain; but, according to Kessler,§ it does not occur within the limits of the Russian empire.

* See Bell. "British Stalk-eyed Crustacea," p. 237.

† Since the statement respecting the occurrence of crayfishes in Spain on p. 44 was printed.

‡ "Die Crustaceen des Südlichen Europas," 1863.

§ "Die Russischen Flusskrebse." Bulletin de la Société Impériale des Naturalistes de Moscow, 1874.

Astacus torrentium appears to be particularly addicted to rapid highland streams and the turbid pools which they feed.

Astacus nobilis is indigenous to France, Germany, and the Italian peninsula. It is said to be found at Nice and at Barcelona, though I cannot hear of it elsewhere in Spain. Its south-eastern limit appears to be the Lake of Zirknitz, in Carniola, not far from the famous caves of Adelsberg. It is not known in Dalmatia, in Turkey, nor in Greece. In the Russian empire, according to Kessler, this crayfish chiefly inhabits the watershed of the Baltic. The northern limit of its distribution lies between Christianstad, in the Gulf of Bothnia (62° 16′ N), and Serdobol, at the northern end of Lake Ladoga. "Eastward of Lake Ladoga it is found in the Uslanka, a tributary of the Swir. It appears to be the only crayfish which exists in the waters which flow from the south into the Gulf of Finland and into the Baltic; except in those streams and lakes which have been artificially connected with the Volga, and in which it is partially replaced by *A. leptodactylus*." It still inhabits the Lakes of Beresai and Bologoe, as well as the affluents of the Msta and the Wolchow; and it is met with in affluents of the Dnieper, as far as Mohilew. *Astacus nobilis* is also found in Denmark and Southern Sweden; but, in the latter country, its introduction appears to have been artificial. This crayfish is said occasionally to be met with on the Livonian coast in the waters of the Baltic, which, however, it must be remembered, are much less salt than ordinary sea water.

It will be observed that while the two forms, *A. torrentium* and *A. nobilis*, are intermixed over a large part of Central Europe, *A. torrentium* has a wider north-westward, south-westward, and south-eastward extension, being the sole occupant of Britain, and apparently of the greater part of Spain and of Greece. On the other hand, in the northern and eastern parts of Central Europe, *A. nobilis* appears to exist alone.

Further to the east, a new form, *Astacus leptodactylus* (fig. 75), makes its appearance. Whether *A. leptodactylus* exists in the upper waters of the Danube, does not appear, but in the lower Danube and in the Theiss it is the dominant, if not the exclusive, crayfish. From hence it extends through all the rivers which flow into the Black, Azov, and Caspian Seas, from Bessarabia and Podolia on the west, to the Ural mountains on the east. In fact, the natural habitat of this crayfish appears to be the watershed of the Pontocaspian area, excluding that part of the Black Sea which lies southward of the Caucasus on the one hand, and of the mouths of the Danube on the other.*

It is a remarkable circumstance that this crayfish not only thrives in the brackish waters of the estuaries of the rivers which debouche into the Black Sea and the Sea of Azov, but that it is found even in the salter

* These statements rest on the authority of Kessler and Gerstfeldt, in their memoirs already cited.

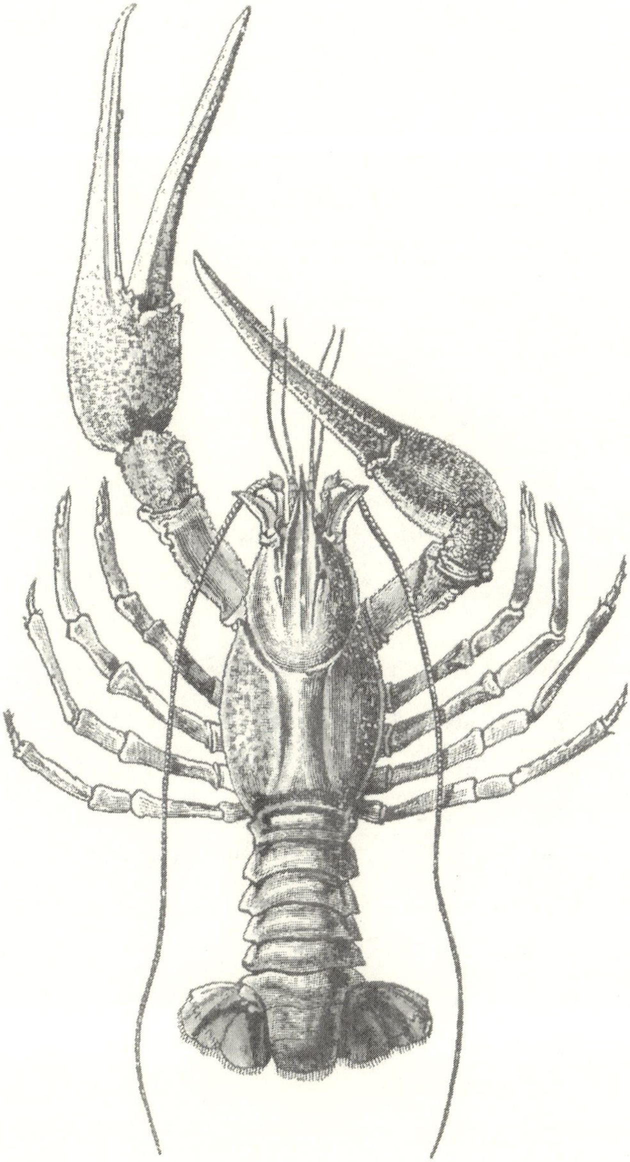

FIG. 75.—*Astacus leptodactylus* (after Rathke, ⅓ nat. size).

southern parts of the Caspian, in which it lives at considerable depths.

In the north, *Astacus leptodactylus* is met with in the rivers which flow into the White Sea, as well as in many streams and lakes about the Gulf of Finland. But it has probably been introduced into these streams by the canals which have been constructed to connect the basin of the Volga with the rivers which flow into the Baltic and into the White Sea. In the latter, the invading *A. leptodactylus* is everywhere overcoming and driving out *A. nobilis* in the struggle for existence, apparently in virtue of its more rapid multiplication.*

In the Caspian and in the brackish waters of the estuaries of the Dniester and the Bug, a somewhat different crayfish, which has been called *Astacus pachypus*, occurs; another closely allied form (*A. angulosus*) is met with in the mountain streams of the Crimea and of the northern face of the Caucasus; and a third, *A. colchicus*, has recently been discovered in the Rion, or Phasis of the ancients, which flows into the eastern extremity of the Black Sea.

With respect to the question whether these Ponto-caspian crayfishes are specifically distinct from one another, and whether the most widely distributed kind, *A. leptodactylus*, is distinct from *A. nobilis*, exactly the same difficulties arise as in the case of the west European

* Kessler (Die Russischen Flusskrebse, l. c. p. 369-70), has an interesting discussion of this question.

crayfishes. Gerstfeldt, who has had the opportunity of examining large series of specimens, concludes that the Pontocaspian crayfishes and *A. nobilis* are all varieties of one species. Kessler, on the contrary, while he admits that *A. angulosus* is, and *A. pachypus* may be, a variety of *A. leptodactylus*, affirms that the latter is specifically distinct from *A. nobilis*.

Undoubtedly, well marked examples of *A. leptodactylus* are very different from *A. nobilis*.

1. The edges of the rostrum are produced into five or six sharp spines, instead of being smooth or slightly serrated as in *A. nobilis*.

2. The fore part of the rostrum has no serrated spinous median keel, such as commonly, though not universally, exists in *A. nobilis*.

3. The posterior end of the post-orbital ridge is still more distinct and spiniform than in *A. nobilis*.

4. The abdominal pleura of *A. leptodactylus* are narrower, more equal sided, and triangular in shape.

5. The chelæ of the forceps, especially in the males, are more elongated; and the moveable and fixed claws are slenderer and have their opposed edges straighter and less tuberculated.

But, in all these respects, individual specimens of *A. nobilis* vary in the direction of *A. leptodactylus* and *vice versâ*; and if *A. angulosus* and *A. pachypus* are varieties of *A. leptodactylus*, I cannot see why Gerstfeldt's conclusion that *A. nobilis* is another variety of the same form need be questioned on morphological grounds. However, Kessler asserts that, in those localities in which *A. leptodactylus* and *A. nobilis* live together, no intermediate forms occur, which is presumptive evidence that they do not intermix by breeding.

No crayfishes are known to inhabit the rivers of the northern Asiatic watershed, such as the Obi, Yenisei, and Lena. None are known * in the sea of Aral, or the great rivers Oxus and Jaxartes, which feed that vast lake; nor any in the lakes of Balkash and Baikal. If further exploration verifies this negative fact, it will be not a little remarkable; inasmuch as two †, if not more, kinds of crayfishes are found in the basin of the great river Amur, which drains a large area of north-eastern Asia, and debouches into the Gulf of Tartary, in about the latitude of York.

Japan has one species (*A. japonicus*), perhaps more; but no crayfish has as yet been made known in any part of eastern Asia, south of Amurland. There are certainly none in Hindostan; none are known in Persia, Arabia, or Syria. In Asia Minor the only recorded locality is the Rion. No crayfish has yet been discovered in the whole continent of Africa.‡

* It would be hazardous, however, to assume that none exist, especially in the Oxus, which formerly flowed into the Caspian.

† *A. dauricus* and *A. Schrenckii*.

‡ Whatever the so-called *Astacus capensis* of the Cape Colony may be, it is certainly not a crayfish.

Thus, on the continent of the old world, the crayfishes are restricted to a zone, the southern limit of which coincides with certain great geographical features; on the west, the Mediterranean, with its continuation, the Black Sea; then the range of the Caucasus, followed by the great Asiatic highlands, as far as the Corea on the east. On the north, though there is no such physical boundary, the crayfishes appear to be entirely excluded from the Siberian river basins; while east and west, though a sea-barrier exists, the crayfishes extend beyond it, to reach the British islands and those of Japan.

Crossing the Pacific, we meet with some half-a-dozen kinds of crayfishes,* different from those of the old world, but still belonging to the genus *Astacus*, in British Columbia, Oregon, and California. Beyond the Rocky Mountains, from the Great Lakes to Guatemala, crayfishes abound, as many as thirty-two different species having been described, but they all belong to the genus *Cambarus* (fig. 63, p. 248). Species of this genus also occur in Cuba,† but, so far as is at present known, not in any of the other West Indian islands. The occurrence of a curious dimorphism among the male *Cambari* has been described by Dr. Hagen; and a blind *Cambarus*

* Dr. Hagen in his "Monograph of the North American Astacidæ," enumerates six species; *A. Gambelii*, *A. klamathensis*, *A. leeniusculus*, *A. nigrescens*, *A. oreganus*, and *A. Trowbridgii*.

† Von Martens. *Cambarus cubensis*. Archiv. für Naturgeschichte, xxxviii.

is found, along with other blind animals, in the subterranean caves of Kentucky.

All the crayfishes of the northern hemisphere belong to the *Potamobiidæ*, and no members of this family are known to exist south of the equator. The crayfishes of the southern hemisphere, in fact, all belong to the division of the *Parastacidæ*, and in respect of the number and variety of forms and the size which they reach, the head-quarters of the *Parastacidæ* is the continent of Australia. Some of the Australian crayfishes (fig. 76) attain a foot or more in length, and are as large as full-sized lobsters. The genus *Engæus* of Tasmania comprises small crayfish which, like some of the *Cambari*, live habitually on land, in burrows which they excavate in the soil.

New Zealand has a peculiar genus of crayfishes, *Paranephrops*, a species of which is found in the Fiji Islands, but none are known to occur elsewhere in Polynesia.

Two kinds of crayfish have been obtained in southern Brazil, and have been described by Dr. v. Martens,* as *A. pilimanus* and *A. brasiliensis*. I have shown that they belong to a peculiar genus, *Parastacus*. The former was procured at Porto Alegre, which is situated in 30° S. Latitude, close to the mouth of the Jacuhy, at the north end of the great Laguna do Patos, which communi-

* Südbrasilische Süss- und Brackwasser Crustaceen, nach den Sammlungen des Dr. Reinh. Hensel. Archiv. für Naturgeschichte, xxxv. 1869.

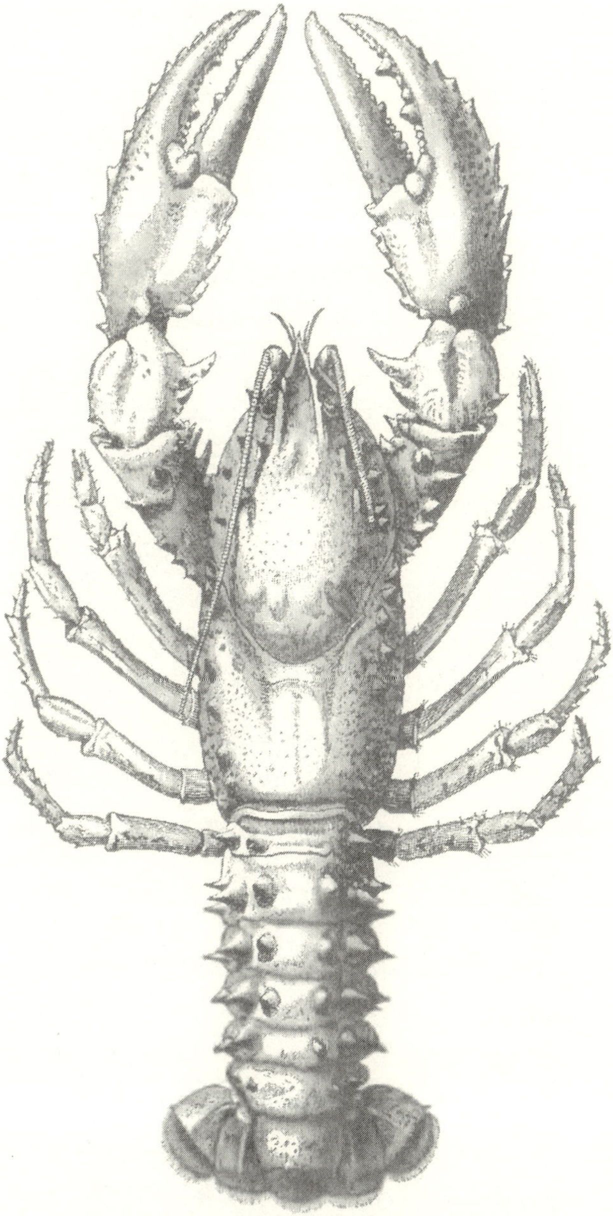

FIG. 76.—Australian Crayfish ($\frac{1}{3}$ nat. size).*

* The nomenclature of the Australian crayfishes requires thorough revision. I therefore, for the present, assign no name to this cray-

cates by a narrow passage with the sea; and also at Sta. Cruz in the upper basin of the Rio Pardo, an affluent of the Jacuhy, "by digging it out of holes in the ground." The latter (*P. brasiliensis*, fig. 64) was obtained at Porto Alegre, and further inland, in the region of the primitive forest at Rodersburg, in shallow streams.

In addition to these, no crayfish have as yet been found in any of the great rivers, such as the Orinoko; the Amazon, in which they were specially sought for by Agassiz; or in the La Plata, on the eastern side of the Andes. But, on the west, an "*Astacus*" *chilensis* is described in the "Histoire Naturelle des Crustacées," (vol. ii. p. 333). It is here stated that this crayfish "habite les côtes du Chili," but the freshwaters of the Chilian coast are doubtless to be understood.

Finally, Madagascar has a genus and species of crayfish (*Astacoides madagascariensis*, fig. 65) peculiar to itself.

On comparing the results obtained by the study of the geographical distribution of the crayfishes with those brought to light by the examination of their morphological characters, the important fact that there is a broad and general correspondence between the two becomes apparent. The wide equatorial belt of the earth's surface which separates the crayfishes of the northern from those of the southern hemisphere, is a sort of geographical

fish. It is probably identical with the *A. nobilis* of Dana and the *A. armatus* of Von Martens.

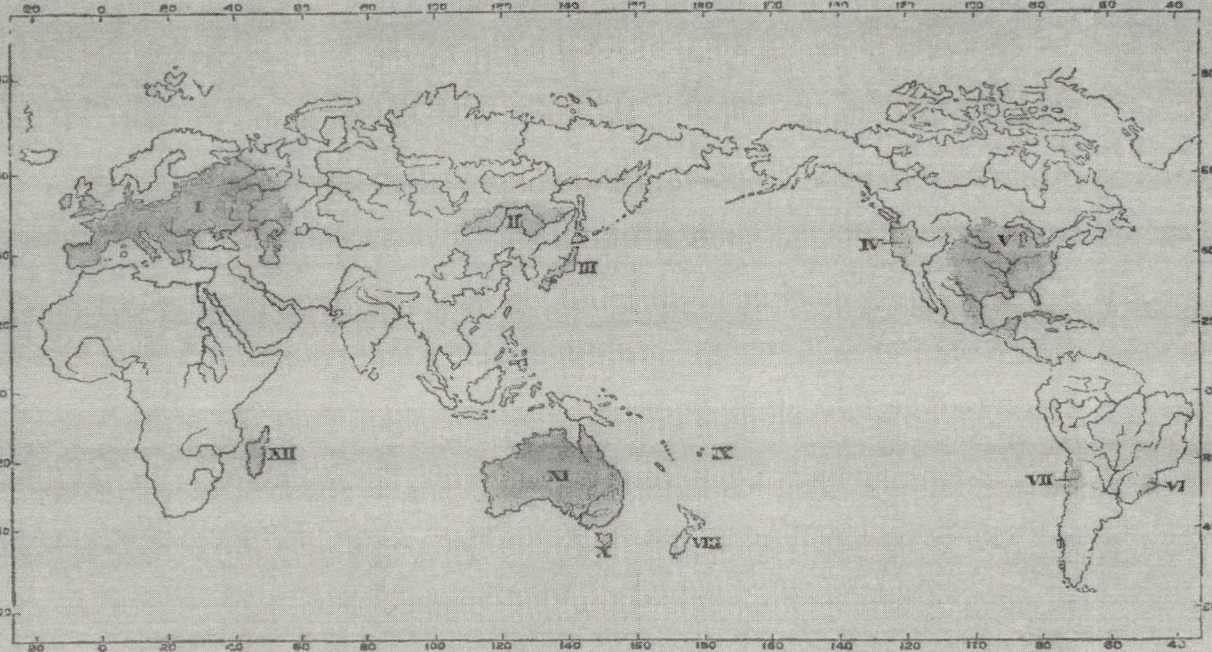

FIG 77.—MAP OF THE WORLD, showing the geographical distribution of the Crayfishes. I. Eur-asiatic Crayfishes; II. Amurland Crayfishes; III. Japanese Crayfishes; IV. Western North American Crayfishes; V. Eastern North American Crayfishes; VI. Brazilian Crayfishes; VII. Chilian Crayfishes; VIII. Novozelanian Crayfishes; IX. Fijian Crayfishes; X. Tasmanian Crayfishes; XI. Australian Crayfishes; XII. Mascarene Crayfishes.

representation of the broad morphological differences which mark off the *Potamobiidæ* from the *Parastacidæ*. Each group occupies a definite area of the earth's surface, and the two are separated by an extensive border-land untenanted by crayfishes.

A similar correspondence is exhibited, though less distinctly, when we consider the distribution of the genera and species of each group. Thus, among the *Potamobiidæ*, *Astacus torrentium* and *nobilis* belong essentially to the northern, western, and southern watersheds of the central European highlands, the streams of which flow respectively into the Baltic and the North Seas, the Atlantic and the Mediterranean (fig. 77, I.); *A. leptodactylus*, *pachypus*, *angulosus*, and *colchicus*, appertain to the Pontocaspian watershed, the rivers of which drain into the Black Sea and the Caspian (I.); while *Astacus dauricus* and *A. Schrenckii* are restricted to the widely separated basin of the Amur, which sheds its waters into the Pacific (II.). The *Astaci* of the rivers of western North America, which flow into the Pacific (IV.), and the *Cambari* of the Eastern or Atlantic water-shed (V.) are separated by the great physical barrier of the Rocky Mountain ranges. Finally, with regard to the *Parastacidæ*, the widely separated geographical regions of New Zealand (VIII.), Australia (IX.), Madagascar (XII.), and South America (VI. and VII.), are inhabited by generically distinct groups.

But when we look more closely into the matter, it will

be found that the parallel between the geographical and the morphological facts cannot be quite strictly carried out.

Astacus torrentium, as we have seen, inhabits both the British Islands and the continent of Europe; nevertheless, there is every reason to believe that twenty miles of sea water is an insuperable barrier to the passage of crayfishes from one land to the other. For though some crayfishes live in brackish water, there is no evidence that any existing species can maintain themselves in the sea. A fact of the same character meets us at the other side of the Eurasiatic continent, the Japanese and the Amurland crayfishes being closely allied; although it is not clear that there are any identical species on the two sides of the Sea of Japan.

Another circumstance is still more remarkable. The West American crayfishes are but little more different from the Pontocaspian crayfishes, than these are from *Astacus torrentium.* On the face of the matter, one might therefore expect the Amurland and Japanese crayfishes, which are intermediate in geographical position, to be also intermediate, morphologically, between the Pontocaspian and the West American forms. But this is not the case. The branchial system of the Amurland *Astaci* appears to be the same as that of the rest of the genus; but, in the males, the third joint (ischiopodite) of the second and third pair of ambulatory limbs is provided with a conical, recurved, hook-like process; while, in the females, the hinder edge of the penultimate thoracic sternum is elevated into a transverse prominence, on the posterior face of which there is a pit or depression.*

In both these characters, but more especially in the former, the Amurland and Japanese *Astaci* depart from both the Pontocaspian and the West American *Astaci*, and approach the *Cambari* of Eastern North America. In these crayfishes, in fact, one or both of the same pairs of legs in the male are provided with similar

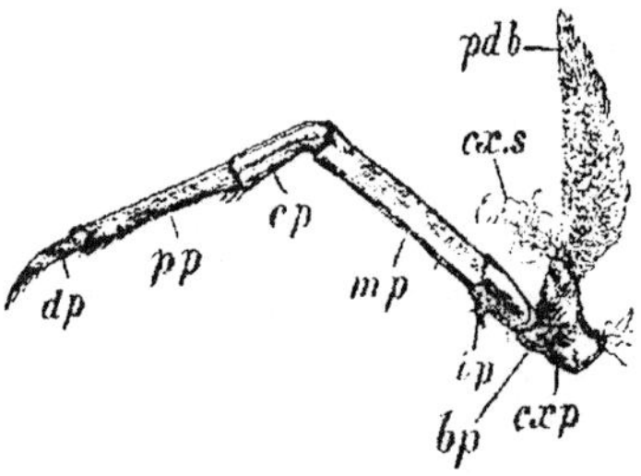

FIG. 78.—*Cambarus* (Guatemala) penultimate leg. *cxp*, coxopodite; *cx.s*, coxopoditic setæ; *pdb*, podobranchia; *bp*, basipodite; *ip*, ischiopodite; *mp*, meropodite; *cp* carpopodite; *pp*. propodite; *dp*, dactylopodite.

hook-like processes; while, in the females, the modification of the penultimate thoracic sternum is carried still further and gives rise to the curious structure described by Dr. Hagen as the "annulus ventralis."

In all the *Cambari*, the pleurobranchiæ appear to be entirely suppressed, and the hindermost podobranchia has no lamina; while the areola is usually extremely narrow. The proportional size of the areola in the Amurland

* Kessler, l. c.

crayfishes is not recorded; in the Japanese crayfish, judging by the figure given by De Haan, it is about the same as in the western *Astaci*. On the other hand, in the West American crayfishes it is distinctly smaller; so that, in this respect, they perhaps more nearly approach the *Cambari*. Unfortunately, nothing is known as to the branchiæ of the Amurland crayfishes. According to De Haan, those of the Japanese species resemble those of the western *Astaci:* as those of the West American *Astaci* certainly do.

With respect to the *Parastacidæ;* in the remarkable length and flatness of the epistoma, the crayfishes of Australia, Madagascar, and South America, resemble one another. But in its peculiar truncated rostrum (see fig. 65) and in the extreme modification of its branchial system, which I have described elsewhere, the Madagascar genus stands alone.

The *Paranephrops* of New Zealand and the Fijis, with its wide and short epistoma, long rostrum, and large antennary squames, is much more unlike the Australian forms than might be expected from its geographical position. On the other hand, considering their wide separation by sea, the amount of resemblance between the New Zealand and the Fiji species is very remarkable.

If the distribution of the crayfishes is compared with that of terrestrial animals in general, the points of difference are at least as remarkable as the resemblances.

With respect to the latter, the area occupied by the *Potamobiidæ*, corresponds roughly with the Palæarctic and Nearctic divisions of the great Arctogæal provinces of distribution indicated by mammals and birds; while distinct groups of crayfishes occupy a larger or smaller part of the other, namely, the Austro-Columbian, Australian, and Novozelanian primary distributional provinces of mammals and birds. Again, the peculiar crayfishes of Madagascar answer to the special features of the rest of the fauna of that island.

But the North American crayfishes extend much further South than the limits of the Nearctic fauna in general; while the absence of any group of crayfishes in Africa, or in the rest of the old world, south of the great Asiatic table-land, forms a strong contrast to the general resemblance of the North African and Indian fauna to that of the rest of Arctogæa. Again, there is no such vast difference between the crayfishes of New Zealand, Australia, and South America, as there is between the mammals and the birds of those regions.

It may be concluded, therefore, that the conditions which have determined the distribution of crayfishes have been very different from those which have governed the distribution of mammals and birds. But if we compare with the distribution of the crayfishes, not that of terrestrial animals in general, but only that of freshwater fishes, some very curious points of approximation become manifest. The *Salmonidæ*, or fishes of the salmon and trout kind, a few of which are exclusively marine, many both marine and freshwater, while others are confined to fresh water, are distributed over the northern hemisphere, in a manner which recalls the distribution of the Potamobine crayfishes,* though they do not extend so far to the South in the new world, while they go a little further, namely, as far as Algeria, Northern Asia Minor, and Armenia, in the old world. With the exception of the single genus *Retropinna*, which inhabits New Zealand, no true salmonoid fish occurs south of the equator; but, as Dr. Günther has pointed out, two groups of freshwater fishes, the *Haplochitonidæ* and the *Galaxidæ*, which stand in somewhat the same relation to the *Salmonidæ* as the *Parastacidæ* do to the *Potamobiidæ*, take the place of the *Salmonidæ* in the fresh waters of New Zealand, Australia, and South America. There are two species of *Haplochiton* in Tierra del Fuego; and of the closely allied genus *Prototroctes*, one species is found in South Australia, and one in New Zealand; of the *Galaxidæ*, the same species, *Galaxias attennuatus*, occurs in the streams of New Zealand, Tasmania, the Falkland Islands, and Peru.

Thus, these fish avoid South Africa, as the crayfishes

* According to Dr. Günther their southern range is similarly limited by the Asiatic Highlands. But they abound in the rivers both of the old and new worlds which flow into the Arctic sea; and though those on

do; but I am not aware that any member of the group is found in Madagascar, and thus completes the analogy.

The preservation of the soft parts of animals in the fossil state depends upon favourable conditions of rare occurrence; and, in the case of the *Crustacea*, it is not often that one can hope to meet with such small hard parts as the abdominal members, in a good state of preservation. But without recourse to the branchial apparatus, and to the abdominal appendages, it might be very difficult to say whether a given crustacean belonged to the Astacine, or to the closely allied Homarine group. Of course, if the accompanying fossils indicated that the deposit in which the remains occur, was of freshwater origin, the presumption in favour of their Astacine nature would be very strong; but if they were inhabitants of the sea, the problem whether the crustacean in question was a marine Astacine, or a true Homarine, might be very hard to solve.

Undoubted remains of crayfishes have hitherto been discovered only in freshwater strata of late tertiary age. In Idaho, North America, Professor Cope * found, in association with *Mastodon mirificus*, and *Equus excelsus*, several species, which he considers to be distinct from

the western side of the Rocky Mountains are different from the Eastern American forms, yet there are species common to both the Asiatic and the American coasts of the North Pacific.

* On three extinct *Astaci* from the freshwater Tertiary of Idaho. Proceedings of the American Philosophical Society, 1869–70.

the existing American crayfishes; whether they are *Cambari* or *Astaci* does not appear. But, in the lower chalk of Ochtrup, in Westphalia, and therefore in a marine deposit, Von der Marck and Schlüter * have obtained a single, somewhat imperfect, specimen of a crustacean, which they term *Astacus politus*, and which, singularly enough, has the divided telson found only in the genus *Astacus*. It would be very desirable to know more about this interesting fossil. For the present it affords a strong presumption that a marine Potamobine existed as far back as the earlier part of the cretaceous epoch.

Such are the more important facts of Morphology, Physiology, and Distribution, which make up the sum of our present knowledge of the Biology of Crayfishes. The imperfection of that knowledge, especially as regards the relations between Morphology and Distribution, becomes a serious drawback when we attack the final problem of Biology, which is to find out why animals of such structure and active powers, and so localized, exist?

It would appear difficult to frame more than two fundamental hypotheses in attempting to solve this problem. Either we must seek the origin of crayfishes in conditions extraneous to the ordinary course of natural

* Neue Fische und Krebse aus der Kreide von Westphalen. Palæontographica, Bd. XV., p. 302; tab. XLIV., figs. 4 and 5.

operations, by what is commonly termed Creation; or we must seek for it in conditions afforded by the usual course of nature, when the hypothesis assumes some shape of the doctrine of Evolution. And there are two forms of the latter hypothesis; for, it may be assumed, on the one hand, that crayfishes have come into existence, independently of any other form of living matter, which is the hypothesis of spontaneous or equivocal generation, or abiogenesis; or, on the other hand, we may suppose that crayfishes have resulted from the modification of some other form of living matter; and this is what, to borrow a useful word from the French language, is known as *transformism.*

I do not think that any hypothesis respecting the origin of crayfishes can be suggested, which is not referable to one or other of these, or to a combination of them.

As regards the hypothesis of creation, little need be said. From a scientific point of view, the adoption of this speculation is the same thing as an admission that the problem is not susceptible of solution. Moreover, the proposition that a given thing has been created, whether true or false, is not capable of proof. By the nature of the case direct evidence of the fact is not obtainable. The only indirect evidence is such as amounts to proof that natural agencies are incompetent to cause the existence of the thing in question. But such evidence is out of our reach. The most that can be proved, in any case, is that no known natural cause is competent to produce a given effect; and it is an obvious blunder to confound the demonstration of our own ignorance with a proof of the impotence of natural causes. However, apart from the philosophical worthlessness of the hypothesis of creation, it would be a waste of time to discuss a view which no one upholds. And, unless I am greatly mistaken, at the present day, no one possessed of knowledge sufficient to give his opinion importance is prepared to maintain that the ancestors of the various species of crayfish were fabricated out of inorganic matter, or brought from nothingness into being, by a creative fiat.

Our only refuge, therefore, appears to be the hypothesis of evolution. And, with respect to the doctrine of abiogenesis, we may also, in view of a proper economy of labour, postpone its discussion until such time as the smallest fragment of evidence that a crayfish can be evolved by natural agencies from not living matter, is brought forward.

In the meanwhile, the hypothesis of transformism remains in possession of the field; and the only profitable inquiry is, how far are the facts susceptible of interpretation, on the hypothesis that all the existing kinds of crayfish are the product of the metamorphosis of other forms of living beings; and that the biological phenomena which they exhibit are the results of the interaction, through past time, of two series of factors: the one, a process of morphological and concomitant physiological modification; the other, a process of change in the condition of the earth's surface.

If we set aside, as not worth serious consideration, the assumption that the *Astacus torrentium* of Britain was originally created apart from the *Astacus torrentium* of the Continent; it follows, either that this crayfish has passed across the sea by voluntary or involuntary migration; or that the *Astacus torrentium* existed before the English Channel, and spread into England while these islands were still continuous with the European mainland; and that the present isolation of the English crayfishes from the members of the same species on the Continent is to be accounted for by those changes in the physical geography of western Europe which, as there is abundant evidence to prove, have separated the British Islands from the mainland.

There is no evidence that our crayfish has been purposely introduced by human agency into Great Britain; and from the mode of life of crayfish and the manner in which the eggs are carried about by the parent during their development, transport by birds or floating timber would seem to be out of the question. Again, although *Astacus nobilis* is said to venture into the brackish waters of the Gulf of Finland, and *A. leptodactylus*, as we have seen, makes itself at home in the more or less salt Caspian, there is no reason to believe that *Astacus torrentium* is capable of existing in seawater, still less of crossing the many miles of sea which separate England from even the nearest point of the Continent. In fact, the existence of the same kind of crayfish on both sides of the Channel appears to be only a case of the general truth, that the Fauna of the British Islands is identical with a part of that of the Continent; and as our foxes, badgers, and moles certainly have neither swum across, nor been transported by man, but existed in Britain while it was still continuous with western Europe, and have been isolated by the subsequent intervention of the sea, so we may confidently explain the presence of *Astacus torrentium* by reference to the same operation.

If we take into account the occurrence of *Astacus nobilis* over so large a part of the area occupied by *Astacus torrentium;* its absence in the British Islands, and in Greece; and the closer affinity which exists between *A. nobilis* and *A. leptodactylus*, than between *A. nobilis* and *A. torrentium;* it seems not improbable that *Astacus torrentium* was the original tenant of the whole western European area outside the Ponto-Caspian watershed; and that *A. nobilis* is an invading offshoot of the Ponto-Caspian or *leptodactylus* form which has made its way into the western rivers in the course of the many changes of level which central Europe has undergone; in the same way as *A. leptodactylus* is now passing into the rivers of the Baltic provinces of Russia.

The study of the glacial phenomena of central Europe

has led Sartorius von Waltershausen* to the conclusion that at the time when the glaciers of the Alps had a much greater extension than at present, a vast mass of freshwater extended from the valley of the Danube to that of the Rhone, around the northern escarpment of the Alpine chain, and connected the head-waters of the Danube with those of the Rhine, the Rhone, and the northern Italian rivers. As the Danube debouches into the Black Sea, and this was formerly connected with the Aralo-Caspian Sea, an easy passage would thus be opened up by which crayfishes might pass from the Aralo-Caspian area to western Europe. If they spread by this road, the *Astacus torrentium* may represent the first wave of migration westward, while *A. nobilis* answers to a second, and *A. leptodactylus*, with its varieties, remains as the representative of the old Aralo-Caspian crayfishes. And thus the crayfishes would present a curious parallel with the Iberian, Aryan, and Mongoloid streams of westward movement among mankind.

If we thus suppose the western Eurasiatic crayfishes to be simply varieties of a primitive Aralo-Caspian stock, their limitation to the south by the Mediterranean and by the great Asiatic highlands becomes easily intelligible.

The extremely severe climatal conditions which obtain in northern Siberia may sufficiently account for the

* "Untersuchungen ueber die Klimate der Gegenwart und der Vorwelt." Natuurkundige Verhandelingen van de Hollandsche Maatschappij der Wetenschappen te Haarlem, 1865.

absence of crayfishes (if they are really absent) in the rivers Obi, Yenisei, and Lena, and in the great lake Baikal, which lies more than 1,300 feet above the sea, and is frozen over from November to May. Moreover, there can be no doubt that, at a comparatively recent period, the whole of this region, from the Baltic to the mouth of the Lena, was submerged beneath a southward extension of the waters of the Arctic ocean to the Aralo-Caspian Sea and Lake Baikal, and a westward extension to the Gulf of Finland.

The great lakes and inland seas which stretch, at intervals, from Baikal, on the east, to Wenner in Sweden, on the west, are simply pools, isolated partly by the rising of the ancient sea-bottom and partly by evaporation; and often completely converted into fresh water by the inflow of the surrounding land-drainage. But the population of these pools was originally the same as that of the Northern Ocean, and a few species of marine crustaceans, mollusks, and fish, besides seals, remain in them as living evidences of the great change which has taken place. The same process which, as we shall see, has isolated the *Mysis* of the Arctic seas in the lakes of Sweden and Finland, has shut up with it other arctic marine crustacea, such as species of *Gammarus* and *Idothea*. And the very same species of *Gammarus* is imprisoned, along with arctic seals, in the waters of Lake Baikal.

The distribution of the American crayfishes agrees equally well with the hypothesis of the northern origin of the stock from which they have been evolved. Even under existing geographical conditions, an affluent of the Mississippi, the St. Peter's river, communicates directly, in rainy weather, with the Red river, which flows into Lake Winnipeg, the southernmost of the long series of intercommunicating lakes and streams, which occupy the low and flat water-parting between the southern and the northern watersheds of the North American Continent. But the northernmost of these, the Great Slave Lake, empties itself by the Mackenzie river into the Arctic Ocean, and thus provides a route by which crayfishes might spread from the north over all parts of North America east of the Rocky Mountains.

The so-called Rocky Mountain range is, in reality, an immense table-land, the edges of which are fringed by two principal lines of mountainous elevations. The table-land itself occupies the place of a great north and south depression which, in the cretaceous epoch, was occupied by the sea and probably communicated with the ocean at its northern, as well as at its southern end. During and since this epoch it became gradually filled up, and it now contains an immense thickness of deposits of all ages from the cretaceous to the pliocene—the earlier marine, the later more and more completely freshwater. During the tertiary epoch, various portions of this area have been occupied by vast lakes, the more northern of which doubtless had outlets into the Northern sea. That crayfish existed in the vicinity of the Rocky Mountains in the latter part of the tertiary epoch is testified by the Idaho fossils. And there is thus no difficulty in understanding their presence in the rivers which have now cut their way to the Pacific coast.

The similarity of the crayfish of the Amurland and of Japan is a fact of the same order as the identity of the English crayfish with the *Astacus torrentium* of the European Continent, and is to be explained in an analogous fashion. For there can be no doubt that the Asiatic continent formerly extended much further to the eastward than it does at present, and included what are now the islands of Japan. Even with this alteration of the geographical conditions, however, it is not easy to see how crayfishes can have got into the Amur-Japanese fresh waters. For a north-eastern prolongation of the Asiatic highlands, which ends to the north in the Stanovoi range, shuts in the Amur basin on the west; while the Amur debouches into the sea of Okhotsk, and the Pacific ocean washes the shores of the Japanese islands.

But there are many grounds for the conclusion that, in the latter half of the tertiary epoch, eastern Asia and North America were connected, and that the chain of the Kurile and Aleutian islands may indicate the position of a great extent of submerged land. In that case, the sea of Okhotsk and Behring's sea may occupy the site of inland waters which formerly placed the mouth of the Amur in direct communication with the Northern Ocean, just as the Black Sea, at present, brings the basin of the

Danube into connection, first with the Mediterranean and then with the western Atlantic; and, as in former times, it gave access from the south to the vast area now drained by the Volga. When the Black Sea communicated with the Aralo-Caspian sea, and this opened to the north into the Arctic sea, a chain of great inland waters must have skirted the eastern frontier of Europe, just such as would now lie on the eastern frontier of Asia if the present coast underwent elevation.

Supposing, however, that the ancestral forms of the *Potamobiidæ* obtained access to the river basins in which they are now found, from the north, the hypothesis that a mass of fresh water once occupied a great part of the region which is now Siberia and the Arctic Ocean, would be hardly tenable, and it is, in fact, wholly unnecessary for our present purpose.

The vast majority of the stalk-eyed crustaceans are, and always have been, exclusively marine animals; the crayfishes, the *Atyidæ*, and the fluviatile crabs (*Thelphusidæ*), being the only considerable groups among them which habitually confine themselves to fresh waters. But even in such a genus as *Penæus*, most of the species of which are exclusively marine, some, such as *Penæus brasiliensis*, ascend rivers for long distances. Moreover, there are cases in which it cannot be doubted that the descendants of marine *Crustacea* have gradually accustomed themselves to fresh water conditions, and have, at the same time, become more or less modified, so that they are no longer absolutely identical with those descendants of their ancestors which have continued to live in the sea.*

In several of the lakes of Norway, Sweden and Finland, and in Lake Ladoga, in Northern Europe; in Lake Superior and Lake Michigan, in North America; a small crustacean, *Mysis relicta*, occurs in such abundance as to furnish a great part of the supply of food to the fresh water fishes which inhabit these lakes. Now, this *Mysis relicta* is hardly distinguishable from the *Mysis oculata* which inhabits the Arctic seas, and is certainly nothing but a slight variety of that species.

In the case of the lakes of Norway and Sweden, there is independent evidence that they formerly communicated with the Baltic, and were, in fact, fiords or arms of the sea. The communication of these fiords with the sea having been gradually cut off, the marine animals they contained have been imprisoned; and as the water has been slowly changed from salt to fresh by the drainage of the surrounding land, only those which were able to withstand the altered conditions have survived. Among these is the *Mysis oculata*, which has in the meanwhile undergone the slight variation which has converted it into *Mysis relicta*. Whether the same explanation ap-

* See on this interesting subject: Martens, "On the occurrence of marine animal forms in fresh water." Annals of Natural History, 1858: Lovèn. "Ueber einige im Wetter und Wener See gefundene Crustaceen." Halle Zeitschrift für die Gesammten Wissenschaften, xix., 1862: G. O. Sars, "Histoire Naturelle des Crustacés d'eau douce de Norvège," 1867.

plies to Lakes Superior and Michigan, or whether the *Mysis oculata* has not passed into these masses of fresh water by channels of communication with the Arctic Ocean which no longer exist, is a secondary question. The fact remains that *Mysis relicta* is a primitively marine animal which has become completely adapted to fresh-water life.

Several species of prawns (*Palæmon*) abound in our own seas. Other marine prawns are found on the coasts of North America, in the Mediterranean, in the South Atlantic and Indian Oceans, and in the Pacific as far south as New Zealand. But species of the same genus (*Palæmon*) are met with, living altogether in fresh water, in Lake Erie, in the rivers of Florida, in the Ohio, in the rivers of the Gulf of Mexico, of the West India Islands and of eastern South America, as far as southern Brazil, if not further; in those of Chili and those of Costa Rica in western South America; in the Upper Nile, in West Africa, in Natal, in the Islands of Johanna, Mauritius, and Bourbon, in the Ganges, in the Molucca and Philippine Islands, and probably elsewhere.

Many of these fluviatile prawns differ from the marine species not only in their great size (some attaining a foot or more in length), but still more remarkably in the vast development of the fifth pair of thoracic appendages. These are always larger than the slender fourth pair (which answer to the forceps of the crayfishes); and, in the m les especially, they are very long and strong, and are terminated by great chelæ, not unlike those of the crayfishes. Hence these fluviatile prawns (known in many places by the name of "Cammarons") are not unfrequently confounded with true crayfishes; though

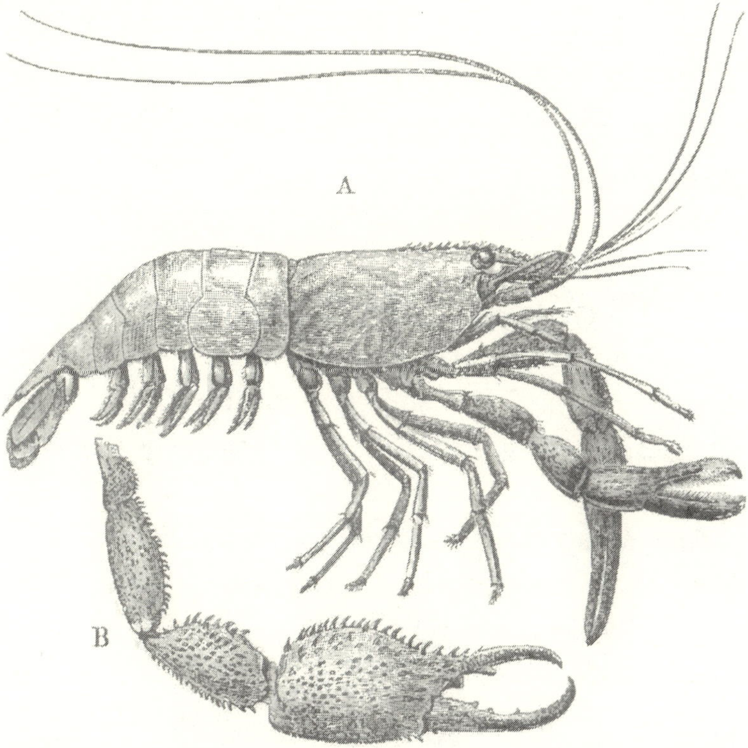

FIG. 79. *Palæmon jamaicensis* (about $\frac{5}{7}$ nat. size). A, female; B, fifth thoracic appendage of male.

the fact that there are only three pair of ordinary legs behind the largest, forceps-like pair, is sufficient at once to distinguish them from any of the *Astacidæ*.

Species of these large-clawed prawns live in the

brackish water lagoons of the Gulf of Mexico, but I am not aware that any of them have yet been met with in the sea itself. The *Palæmon lacustris* (*Anchistia migratoria*, Heller) abounds in fresh-water ditches and canals between Padua and Venice, and in the Lago di Garda, as well as in the brooks of Dalmatia; but its occurrence in the Adriatic or the Mediterranean, which has been asserted, appears to be doubtful. So the Nile prawn, though very similar to some Mediterranean prawns, does not seem to be identical with any at present known.*

In all these cases, it appears reasonable to apply the analogy of the *Mysis relicta*, and to suppose that the fluviatile prawns are simply the result of the adaptive modification of species which, like their congeners, were primitively marine.

But if the existing sea prawns were to die out, or to be beaten in the struggle for existence, we should have, scattered over the world in isolated river basins, more or less distinct species of freshwater prawns,† the areas inhabited by which might hereafter be indefinitely enlarged or diminished, by alteration in the elevation of the

* Heller, "Die Crustaceen des südlichen Europas," p. 259. Klunzinger, "Ueber eine Süsswasser-crustacee im Nil," with the notes by von Martens and von Siebold: Zeitschrift für Wissenschaftliche Zoologie, 1866.

† This seems actually to have happened in the case of the widely-spread allies and companions of the fluviatile prawns, *Atya* and *Caridina*. I am not aware that truly marine species of these genera are known.

land and by other changes in physical geography. And, indeed, under these circumstances, the freshwater prawns themselves might become so much modified, that, even if the descendants of their ancestors remained unchanged in structure and habits in the sea, the relationship of the two might no longer be obvious.

These considerations appear to me to indicate the direction in which we must look for a rational explanation of the origin of crayfishes and their present distribution.

I have no doubt that they are derived from ancestors which lived altogether in the sea, as the great majority of the *Mysidæ* and many of the prawns do now; and that, of these ancestral crayfishes, there were some which, like *Mysis oculata* or *Penæus brasiliensis*, readily adapted themselves to fresh water conditions, ascended rivers, and took possession of lakes. These, more or less modified, have given rise to the existing crayfishes, while the primitive stock would seem to have vanished. At any rate, at the present time, no marine crustacean with the characters of the *Astacidæ* is known.

As crayfishes have been found in the later tertiaries of North America, we shall hardly err in dating the existence of these marine crayfishes at least as far back as the miocene epoch; and I am disposed to think that, during the earlier tertiary and later mesozoic periods, these *Crustacea* not only had as wide a distribution as the Prawns and *Penæi* have now, but were differentiated into two groups, one with the general characters of the

Potamobiidæ in the northern hemisphere, and another, with those of the *Parastacidæ*, in the southern hemisphere.

The ancestral Potamobine form probably presented the peculiarities of the *Potamobiidæ* in a less marked degree than any existing species does. Probably the four pleurobranchiæ were all equally well developed; the laminæ of the podobranchiæ smaller and less distinct from the stem; the first and second abdominal appendages less specialised; and the telson less distinctly divided. So far as the type was less specially Potamobine, it must have approached the common form in which *Homarus* and *Nephrops* originated. And it is to be remarked that these also are exclusively confined to the northern hemisphere.

The wide range and close affinity of the genera *Astacus* and *Cambarus* appear to me to necessitate the supposition that they are derived from some one already specialised Potamobine form; and I have already mentioned the grounds upon which I am disposed to believe that this ancestral Potamobine existed in the sea which lay north of the miocene continent in the northern hemisphere.

In the marine primitive crayfishes south of the equator, the branchial apparatus appears to have suffered less modification, while the suppression of the first abdominal appendages, in both sexes, has its analogue among the *Palinuridæ*, the headquarters of which are in the southern hemisphere. That they should have ascended the rivers of New Zealand, Australia, Madagascar, and South America, and become fresh water *Parastacidæ*, is an assumption which is justified by the analogy of the fresh-water prawns. It remains to be seen whether marine *Parastacidæ* still remain in the South Pacific and Atlantic Oceans, or whether they have become extinct.

In speculating upon the causes of an effect which is the product of several co-operating factors, the nature of each of which has to be divined by reasoning backwards from its effects, the probability of falling into error is very great. And this probability is enhanced when, as in the present case, the effect in question consists of a multitude of phenomena of structure and distribution about which much is yet imperfectly known. Hence the preceding discussion must rather be regarded as an illustration of the sort of argumentation by which a completely satisfactory theory of the ætiology of the crayfish will some day be established, than as sufficing to construct such a theory. It must be admitted that it does not account for the whole of the positive facts which have been ascertained; and that it requires supplementing, in order to furnish even a plausible explanation of various negative facts.

The positive fact which presents a difficulty is the closer resemblance between the Amur-Japanese crayfish and the East American *Cambari*, than between the latter and the West American *Astaci;* and the closer resemblance between the latter and the Pontocaspian crayfish, than either bear to the Amur-Japanese form. If the facts had been the other way, and the West American and Amur-Japanese crayfish had changed places, the case would have been intelligible enough. The primitive Potamobine stock might then have been supposed to have differentiated itself into a western astacoid, and an eastern cambaroid form;* the latter would have ascended the American, and the former the Asiatic rivers. As the matter stands, I do not see that any plausible explanation can be offered without recourse to suppositions respecting a former more direct communication between the mouth of the Amur, and that of the North American rivers, in favour of which no definite evidence can be offered at present.

The most important negative fact which remains to be accounted for is the absence of crayfishes in the rivers of a large moiety of the continental lands, and in numerous islands. Differences of climatal conditions are obviously inadequate to account for the absence of crayfishes in Jamaica, when they are present in Cuba; for their absence in Mozambique, and the islands of Johanna and Mauritius, when they are present in Madagascar; and for their absence in the Nile, when they exist in Guatemala.

* Just as there is an American form of *Idothea* and an Asiatic form in the Arctic ocean at the present day.

At present, I confess that I do not see my way to a perfectly satisfactory explanation of the absence of crayfishes in so many parts of the world in which they might, *à priori*, be expected to exist; and I can only suggest the directions in which an explanation may be sought.

The first of these is the existence of physical obstacles to the spread of crayfishes, at the time at which the Potamobine and the Parastacine stocks respectively began to take possession of the rivers, some of which have now ceased to exist; and the second is the probability that, in many rivers which have been accessible to crayfishes, the ground was already held by more powerful competitors.

If the ancestors of the Potamobine crayfishes originated only among those primitive crayfishes which inhabited the seas north of the miocene continent, their present limitation to the south, in the old world, is as easily intelligible as is their extension southward, in the course of the river basins of Northern America as far as Guatemala, but no further. For the elevation of the Eurasiatic highlands had commenced in the miocene epoch, while the isthmus of Panama was interrupted by the sea.

With respect to the Southern hemisphere, the absence of crayfishes in Mauritius and in the islands of the Indian Ocean, though they occur in Madagascar, may be due to the fact that the former islands are of comparatively late volcanic origin; while Madagascar is the remnant of a very ancient continental area, the oldest indigenous population of which, in all probability, is directly descended from that which occupied it at the beginning of the tertiary epoch. If Parastacine *Crustacea* inhabited the southern hemisphere at this period, and subsequently became extinct as marine animals, their preservation in the freshwaters of Australia, New Zealand, and the older portions of South America may be understood. The difficulty of the absence of crayfishes in South Africa * remains; and all that can be said is, that it is a difficulty of the same nature as that which confronts us when we compare the fauna of South Africa in general with that of Madagascar. The population of the latter region has a more ancient aspect than that of the former; and it may be that South Africa, in its present shape, is of very much later date than Madagascar.

With respect to the second point for consideration, it is to be remarked that, in the temperate regions of the world, the crayfishes are by far the largest and strongest of any of the inhabitants of freshwater, except the *Vertebrata;* and that while frogs and the like fall an easy prey to them, they must be formidable enemies and competitors even to fishes, aquatic reptiles, and the smaller aquatic mammals. In warm climates, however, not only the large prawns which have been mentioned, but *Atyæ*

* But it must be remembered that we have as yet everything to learn respecting the fauna of the great inland lakes and river systems of South Africa.

and fluviatile crabs (*Thelphusa*) compete for the possession of the freshwaters; and it is not improbable that under some circumstances, they may be more than a match for crayfishes; so that the latter might either be driven out of territory they already occupied, as *Astacus leptodactylus* is driving out *A. nobilis* in the Russian rivers; or might be prevented from entering rivers already tenanted by their rivals.

In connection with this speculation, it is worthy of remark that the area occupied by the fluviatile crabs is very nearly the same as that zone of the earth's surface from which crayfish are excluded, or in which they are scanty. That is to say, they are found in the hotter parts of the eastern side of the two Americas, the West Indies, Africa, Madagascar, Southern Italy, Turkey and Greece, Hindostan, Burmah, China, Japan, and the Sandwich Islands. The large-clawed fluviatile prawns are found in the same regions of America, on both east and west coasts, in Africa, Southern Asia, the Moluccas, and the Philippine Islands; while the *Atyidæ* not only cover the same area, but reach Japan, extend over Polynesia, to the Sandwich Islands, on the north, and New Zealand, on the south, and are found on both shores of the Mediterranean; a blind form (*Troglocaris Schmidtii*), in the Adelsberg caves, representing the blind *Cambarus* of the caves of Kentucky.

The hypothesis respecting the origin of crayfishes

which has been tentatively put forward in the preceding pages, involves the assumption that marine Crustacea of the astacine type were in existence during the deposition of the middle tertiary formations, when the great continents began to assume their present shape. That such was the case there can be no doubt, inasmuch as abundant remains of Crustacea of that type occur still earlier in the mesozoic rocks. They prove the existence of ancient crustaceans, from which the crayfishes may have been derived, at that period of the earth's history when the conformation of the land and sea were such as to admit of their entering the regions in which we now find them.

The materials which have, up to the present, time been collected are too scanty to permit of the tracing out of all the details of the genealogy of the crayfish. Nevertheless, the evidence which exists is perfectly clear, as far as it goes, and is in complete accordance with the requirements of the doctrine of evolution.

Mention has been made of the close affinity between the crayfishes and the lobsters—the *Astacina* and the *Homarina*; and it fortunately happens that these two groups, which may be included under the common name of the *Astacomorpha*, are readily distinguishable from all the other *Podophthalmia* by peculiarities of their exoskeleton which are readily seen in all well-preserved fossils. In all, as in the crayfish, there are large forceps, followed by two pairs of chelate ambulatory limbs, while the succeeding two pairs of legs are terminated by simple claws. The exopodite of the last abdominal appendage is divided into two parts by a transverse suture. The pleura of the second abdominal somite are larger than the others, and overlap those of the first somite, which are very small. Any fossil crustacean which presents all these characters, is certainly one of the *Astacomorpha.*

The *Astacina,* again, are distinguished from the *Homarina* by the mobility of the last thoracic somite, and the characters of the first and second abdominal appendages, when they are present; or by their entire absence. But it is so difficult to make out anything about either of these characters in fossils, that, so far as I am aware, we know nothing about them in any fossil Astacomorph. And hence, it may be impossible to say to which division any given form belongs, unless its resemblances to known types are so minute and so close as to remove doubt.

For the present purpose, the series of the fossiliferous rocks may be grouped as follows:—1. Recent and Quaternary. 2. Newer Tertiary (Pliocene and Miocene). 3. Older Tertiary (Eocene). 4. Cretaceous (Chalk, Greensand and Gault). 5. Wealden. 6. Jurassic (Purbeck to Inferior Oolite). 7. Liassic. 8. Triassic. 9. Permian. 10. Carboniferous. 11. Devonian. 12. Silurian. 13. Cambrian.

Now the oldest known member of the group of the

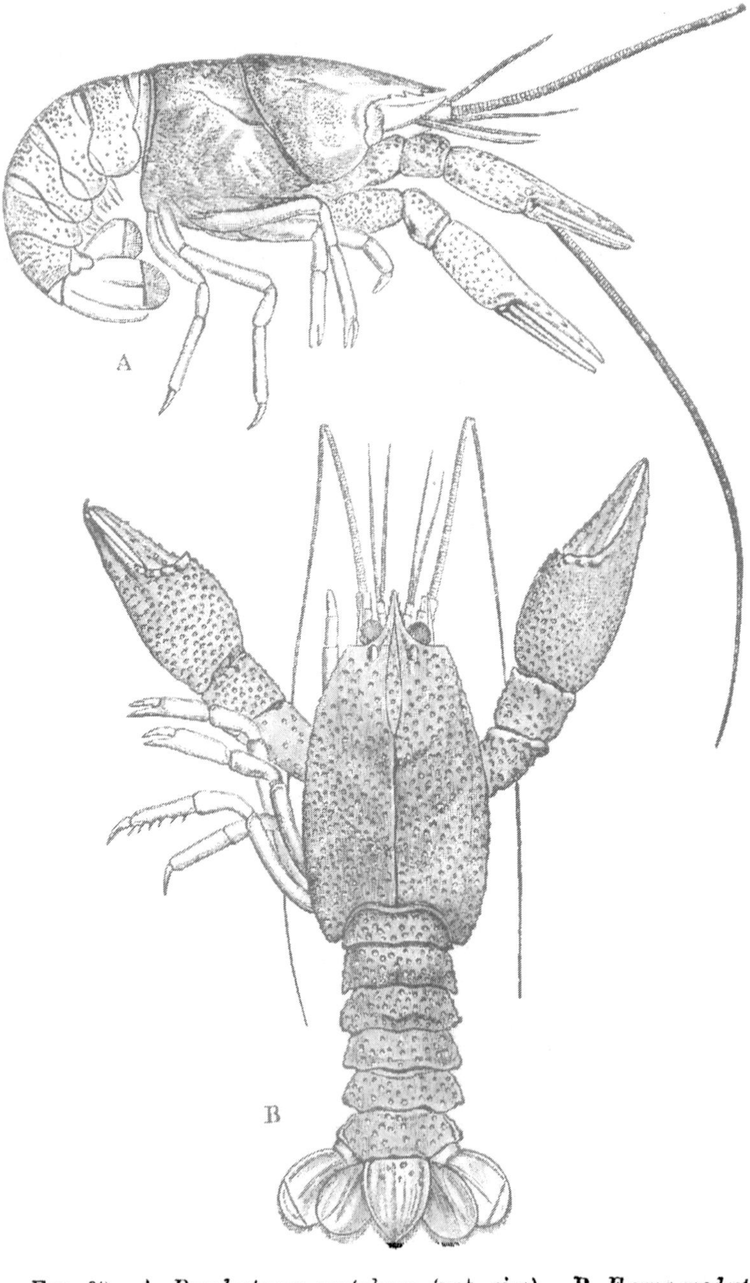

FIG. 80.—A, *Pseudastacus pustulosus* (nat. size). **B,** ***Eryma modestiformis*** (× 2). Both figures after Oppel.

decapod *Podophthalmia* to which the *Astacomorpha* belong occurs in the Carboniferous formation. It is the genus *Anthrapalæmon*—a small and very curious crustacean, about which nothing more need be said at present, as it does not appear to have special affinities with the *Astacomorpha*. In the later formations, up to the top of the Trias, podophthalmatous *Crustacea* are very rare; and, unless the Triassic genus *Pemphix* is an exception, no Astacomorphs are known to occur in them. The specimens of *Pemphix* which I have examined are not sufficiently complete to enable me to express any opinion about them.

The case is altered when we reach the Middle Lias. In fact this yields several forms of a genus, *Eryma* (fig. 80, B), which also occurs in the overlying strata almost up to the top of the Jurassic series, and presents so many variations that nearly forty different species have been recognised. *Eryma* is, in all respects, an Astacomorph, and so far as can be seen, it differs from the existing genera only in such respects as those in which they differ from one another. Thus it is quite certain that Astacomorphous *Crustacea* have existed since a period so remote as the older part of the Mesozoic period; and any hesitation in admitting this singular persistency of type on the part of the crayfishes, is at once removed by the consideration of the fact that, along with *Eryma*, in the Middle Lias, prawn-like *Crustacea*, generically identical with the existing *Penæus*, flourished in the sea and left their remains in the mud of the ancient sea bottom.

Eryma is the only crustacean, which can be certainly ascribed to the *Astacomorpha*, that has hitherto been found in the strata from the Middle Lias to the lithographic slates; which last lie in the upper part of the Jurassic series. In the freshwater beds of the Wealden, no *Astacomorpha* are known, and although no very great

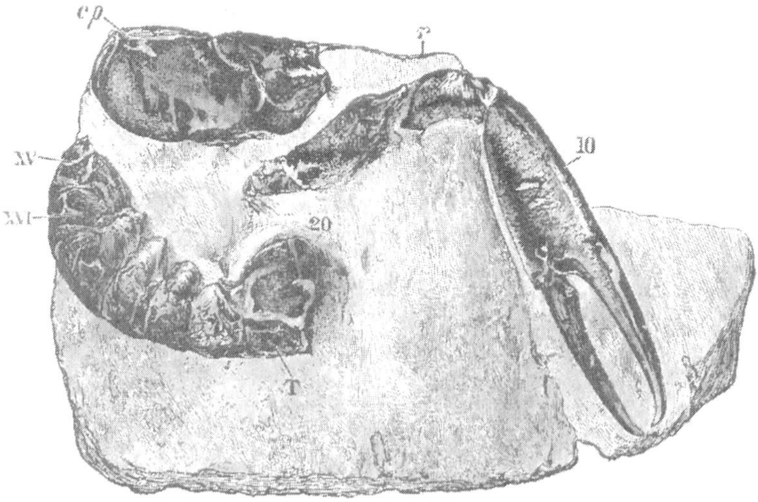

FIG. 81.—*Hoploparia longimana* (⅔ nat. size).—*cp*, carapace; *r*, rostrum, T, telson; XV., XVI., first and second abdominal somites; 10, forceps; 20, last abdominal appendage.

weight is to be attached to a negative fact of this kind, it is, so far, evidence that the *Astacomorpha* had not yet taken to freshwater life. In the marine deposits of the Cretaceous epoch, however, astacomorphous forms, which are known by the generic names of *Hoploparia* and *Enoploclytia*, are abundant.

The differences between these two genera, and between both and *Eryma*, are altogether insignificant from a broad morphological point of view. They appear to me to be of less importance than those which obtain between the different existing genera of crayfishes.

Hoploparia is found in the London clay. It therefore extends beyond the bounds of the Mesozoic epoch into the older Tertiary. But when this genus is compared with the existing *Homarus* and *Nephrops*, it is found partly to resemble the one and partly the other. Thus, on one line, the actual series of forms which have succeeded one another from the Liassic epoch to the present day, is such as must have existed if the common lobster and the Norway lobster are the descendants of *Erymoid* crustaceans which inhabited the seas of the Liassic epoch.

Side by side with *Eryma*, in the lithographic slates, there is a genus, *Pseudastacus* (fig. 80, A), which, as its name implies, has an extraordinarily close resemblance to the crayfishes of the present day. Indeed there is no point of any importance in which (in the absence of any knowledge of the abdominal appendages in the males) it differs from them. On the other hand, in some features, as in the structure of the carapace, it differs from *Eryma*, much as the existing crayfishes differ from *Nephrops*. Thus, in the latter part of the Jurassic epoch, the Astacine type was already distinct from the Homarine type, though both were marine; and, since *Eryma* begins at least as early as the Middle Lias, it is possible that *Pseudastacus* goes back as far, and that the common protastacine form is to be sought in the Trias. *Pseudastacus* is found in the marine cretaceous rocks of the Lebanon, but has not yet been traced into the Tertiary formations.

I am disposed to think that *Pseudastacus* is comparable to such a form as *Astacus nigrescens* rather than to any of the *Parastacidæ*, as I doubt the existence of the latter group at any time in northern latitudes.

In the chalk of Westphalia (also a marine deposit) a single specimen of another Astacomorph has been discovered, which possesses an especial interest as it is a true *Astacus* (*A. politus*, Von der Marck and Schlüter), provided with the characteristic transversely divided telson which is found in the majority of the *Potamobiidæ*.

If we arrange the results of palæontological inquiry which have now been stated in the form of a table such as that which is given on the following page, the significance of the succession of astacomorphous forms, in time, becomes apparent.

SUCCESSIVE FORMS OF THE ASTACOMORPHOUS TYPE.

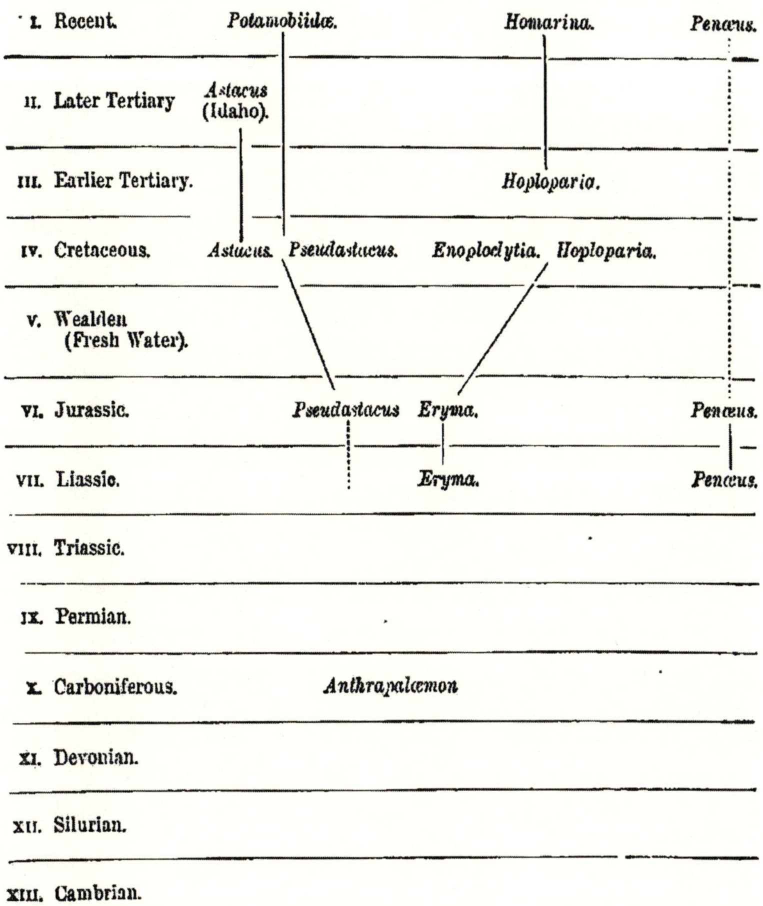

If an Astacomorphous crustacean, having characters intermediate between those of *Eryma* and those of *Pseudastacus*, existed in the Triassic epoch or earlier; if it gradually diverged into Pseudastacine and Erymoid forms; if these again took on Astacine and Homarine characters, and finally ended in the existing *Potamobiidæ* and *Homarina*, the fossil forms left in the track of this process of evolution would be very much what they actually are. Up to the end of the Mesozoic epoch the only known *Potamobiidæ* are marine animals. And we have already seen that the facts of distribution suggest the hypothesis that they must have been so, at least up to this time.

Thus, with respect to the Ætiology of the crayfishes, all the known facts are in harmony with the requirements of the hypothesis that they have been gradually evolved in the course of the Mesozoic and subsequent epochs of the world's history from a primitive Astacomorphous form.

And it is well to reflect that the only alternative supposition is, that these numerous successive and coexistent forms of insignificant animals, the differences of which require careful study for their discrimination, have been separately and independently fabricated, and put into the localities in which we find them. By whatever verbal fog the question at issue may be hidden, this is the real nature of the dilemma presented to us not only by the crayfish, but by every animal and by every plant; from man to the humblest animalcule; from the spreading beech and towering pine to the *Micrococci* which lie at the limit of microscopic visibility.

NOTES.

NOTE I., CHAPTER I., p. 17.

THE CHEMICAL COMPOSITION OF THE EXOSKELETON.

THE harder parts of the exoskeleton of the crayfish contain rather more than half their weight of calcareous salts. Of these nearly seven-eighths consist of carbonate of lime, the rest being phosphate of lime.

The animal matter consists for the most part of a peculiar substance termed *Chitin*, which enters into the composition of the hard parts not only of the *Arthropoda* in general but of many other invertebrated animals. Chitin is not dissolved even by hot caustic alkalies, whence the use of solutions of caustic potash and soda in cleaning the skeletons of crayfishes. It is soluble in cold concentrated hydrochloric acid without change, and may be precipitated from its solution by the addition of water.

Chitin contains nitrogen, and according to the latest investigations (Ledderhose, "Ueber Chitin und seine Spaltungs-produkte:" Zeitschrift für Physiologische Chemié, II. 1879) its composition is represented by the formula $C_{15} H_{26} N_2 O_{10}$.

NOTE II., CHAPTER I., p. 29.

THE CRAB'S EYES, OR GASTROLITHS.

The "Gastroliths," as the "crab's eyes" may be termed, are found fully developed only in the latter part of the summer season, just before ecdysis sets in. They then give rise to rounded prominences, one on each side of the anterior part of the cardiac division of the stomach. The proper wall of the stomach is continued over the outer surface of the prominence ; and, in fact, forms the outer wall of the chamber in which the gastrolith is contained, the inner wall being formed by the cuticular lining of the stomach. When the outer wall is cut through, it is readily detached from the convex outer surface of the gastrolith, with which it is in close contact. The inner surface of the gastrolith is usually flat or slightly concave. Sometimes it is strongly adherent to the chitonous cuticula ; but when fully formed it is readily detached from the latter. Thus the proper wall of the stomach invests only the outer face of the gastrolith, the inner face of which is adherent to, or at any rate in close contact with, the cuticula. The gastrolith is by no means a mere concretion, but is a cuticular growth, having a definite structure. Its inner surface is smooth, but the outer surface is rough, from the projection of irregular ridges which form a kind of meshwork. A vertical section shows that it is composed of thin superimposed layers, of which the inner are parallel with the flat inner surface, while the outer becomes gradually concentric with the outer surface. Moreover, the inner layers are less calcified than the outer, the projections of the outer surface being particularly dense and hard. In fact, the gastroliths are very similar to other hard parts of the exoskeleton in structure, except that the densest layers are nearest the epithelial substratum, instead of furthest away from it.

When ecdysis occurs, the gastroliths are cast off along with the gastric armature in general, into the cavity of the stomach, and are there dissolved, a new cuticle being formed external to them from the proper wall of the stomach. The dissolved calcareous matter is probably used up in the formation of the new exoskeleton.

According to the observations of M. Chantran (Comptes Rendus, LXXVIII. 1874) the gastroliths begin to be formed about forty days before ecdysis takes place in crayfish of four years' old; but the interval is less in younger crayfish, and is not more than ten days during the first year after birth. When shed into the stomach during ecdysis they are ground down, not merely dissolved. The process of destruction and absorption takes twenty-four to thirty hours in very young crayfish, seventy to eighty hours in adults. Unless the gastroliths are normally developed and re-absorbed, ecdysis is not healthily effected, and the crayfish dies in the course of the process.

According to Dulk ("Chemische Untersuchung der Krebsteine:" Müller's Archiv. 1835), the gastroliths have the following composition :—

Animal matter soluble in water	11·43
Animal matter insoluble in water (probably chitin)	4·33
Phosphate of lime	18·60
Carbonate of lime	63·16
Soda reckoned as carbonate	1·41
	98·93

The proportion of mineral to animal matter and of phosphate to carbonate of lime is therefore greater in the gastroliths than in the exoskeleton in general.

NOTE III., CHAPTER I., p. 31.

GROWTH OF CRAYFISH.

The statements in the text, after the words "By the end of the year," regarding the sizes of the crayfish at different ages, are given on the authority of M. Carbonnier (L'Écrevisse. Paris, 1869); but they obviously apply only to the large "Écrevisse à pieds rouges" of France, and not to the English crayfish, which appears to be identical with the "Écrevisse à pieds blancs," and is of much smaller size. According to M. Carbonnier (l. c. p. 51), the young crayfish just born is "un centimètre et demi environ," that is to say, three-fifths of an inch long. The young of the English crayfish still attached to the mother, which I have seen, rarely exceeds half this length.

M. Soubeiran ("Sur l'histoire naturelle et l'education des Écrevisses:" Comptes Rendus, LX. 1865) gives the result of his study of the growth of the crayfishes reared at Clairefontaine, near Rambouillet, in the following table:

	Mean length. Metres.	Mean weight. Grammes.
Crayfish of the year . .	0·025 . .	0·50
" 1 year old . .	0·050 . .	1·50
" 2 years old . .	0·070 . .	3·50
" 3 years " . .	0·090 . .	6·50
" 4 years " . .	0·110 . .	17·50
" 5 years " . .	0·125 . .	18·50
" indeterminate . .	0·1C0 . .	30·00
" very old . .	0·190 . .	125·00

These observations must also apply to the "Écrevisse à pieds rouges."

NOTE IV., CHAPTER I, p. 37.

THE ECDYSES OF CRAYFISHES.

There is a good deal of discrepancy between different observers as to the frequency of the process of ecdysis in crayfishes. In the text I have followed M. Carbonnier, but M. Chantran ("Observations sur l'histoire naturelle des Écrevisses:" Comptes Rendus, LXXI. 1870, and LXXIII. 1871), who appears to have studied the question (on the "écrevisse à picds rouges" apparently) very carefully, declares that the young crayfish moults no fewer than eight times in the course of the first twelve months. The first moult takes place ten days after it is hatched; the second, third, fourth, and fifth, at intervals of from twenty to twenty-five days, so that the young animal moults five times in the course of the ninety to one hundred days of July, August, and September. From the latter month to the end of April in the following year, no ecdysis takes place. The sixth takes place in May, the seventh in June, and the eighth in July. In the second year of its age, the crayfish moults five times, that is to say, in August and in September, and in May, June, and July following. In the third year, the crayfish commonly moults only twice, namely in July and in September. At a greater age than this, the females moult only once a year, from August to September; while the males moult twice, first in June and July; afterwards in August and September.

The details of the process of ecdysis are discussed by Braun, "Ueber die histologischen Vorgänge bei der Häutung von *Astacus fluviatilis.*" Würzburg Arbeiten, Bd. II.

NOTE V., CHAPTER I., p. 39.

REPRODUCTION IN CRAYFISHES.

The males are said to approach the females in November, December, and January, in the case of the French crayfishes. In England they certainly begin as early as the beginning of October, if not earlier. According to M. Chantran (Comptes Rendus, 1870), and M. Gerbe (Comptes Rendus, 1858), the male seizes the female with his pincers, throws her on her back, and deposits the spermatic matter, firstly, on the external plates of the caudal fin; secondly, on the thoracic sterna around the external openings of the oviducts. During this operation, the appendages of the two first abdominal somites are carried backwards, the extremities of the posterior pair are inclosed in the groove of the anterior pair; and the end of the vas deferens becoming everted and prominent, the seminal matter is poured out, and runs slowly along the groove of the anterior appendage to its destination, where it hardens and assumes a vermicular aspect. The filaments of which it is composed are, in fact, tubular spermatophores, and consist of a tough case or sheath filled with seminal matter. The spoon-shaped extremity of the second abdominal appendage, working backwards and forwards in the groove of the anterior appendage, clears the seminal matter out of it, and prevents it from becoming choked.

After an interval which varies from ten to forty-five days, oviposition takes place. The female, resting on her back, bends the end of the abdomen forward over the hinder thoracic sterna, so that a chamber is formed into which the oviducts open. The eggs are passed into the chamber by one operation, usually during the night, and are plunged into a viscous greyish mucus with which it is filled. The spermatozoa pass out of the vermicular spermatophores, and mix with this fluid, in which the peculiarity of their form renders them readily recognisable. The spermatozoa are thus brought into close relation with the ova, but what actually becomes of them is unknown.

The origin of the viscous matter which fills the abdominal chamber when the eggs are deposited in it, and the manner in which these become fixed to the abdominal limbs is discussed by Lereboullet ("Recherches sur le mode de fixation des œufs aux faux pattes abdominaux dans les Écrevisses." Annales des Sciences Naturelles, 4e Ee. T. XIV. 1860), and by Braun (Arbeiten aus dem Zoologisch-Zootomischen Institut in Würzburg, II.).

NOTE VI., CHAPTER I., p. 42.

ATTACHMENT OF THE YOUNG CRAYFISH TO THE MOTHER.

I observe that I had overlooked a passage in the Report on the award of the Prix Montyon for 1872, Comptes Rendus, LXXV. p. 1341, in which M. Chantran is stated to have ascertained that the young crayfishes fix themselves "en saisissant avec un de leurs pinces le filament qui suspend l'œuf à une fausse patte de la mère."

In the paper already cited from the Comptes Rendus for 1870, M. Chantran states that the young remain attached to the mother during ten days after hatching, that is to say, up to the first moult. Detached before this period, they die; but after the first moult, they sometimes leave the mother and return to her again, up to twenty-eight days, when they become independent.

In a note appended to M. Chantran's paper, M. Robin states, that "the young are suspended to the abdomen of the mother by the intermediation of a chitinous hyaline filament, which extends from a point of the internal surface of the shell of the egg as far as the four most internal filaments of each of the lobes of the median membranous plate of the caudal appendage. The filaments exist when the embryos have not yet attained three-fourths of their development." Is this a larval coat? Rathke does not mention it and I have seen nothing of it in those recently hatched young which I have had the opportunity of examining.

NOTE VII., CHAPTER II., p. 64.

THE "SALIVARY" GLANDS AND THE SO-CALLED "LIVER" OF THE CRAYFISH.

Braun (Arbeiten aus dem Zoologisch-Zootomischen Institut in Würzburg, Bd. II. and III.) has described "salivary" glands in the walls of the œsophagus, in the metastoma, and in the first pair of maxillæ of the crayfish.

Hoppe-Seyler (Pflügers Archiv, Bd. XIV. 1877) finds that the yellow fluid ordinarily found in the stomachs of crayfishes always contains peptone. It dissolves fibrin readily, without swelling it up, at ordinary temperatures; more quickly at 40° Centigrade. The action is delayed by even a trace of hydrochloric acid, and is stopped by the addition of a few drops of water containing 0.2 per cent. of that acid. By adding alcohol to the yellow fluid, a precipitate is obtained, which is soluble in water and in glycerine. The aqueous solution of the precipitate has a strong digestive action on fibrin, which is arrested by acidulation with hydrochloric acid. These reactions show that the fluid is very similar to, if not identical with, the pancreatic fluid of vertebrates.

The secretion of the "liver" taken directly from that gland, has a more strongly acid reaction than the fluid in the stomach, but has similar digestive properties. So has an aqueous extract of the gland, and a watery solution of the alcoholic precipitate. The aqueous extract also possesses a strong diastatic action on starch, and breaks up olive oil. There is no more glycogen in the "liver" than is to be found in other organs, and no constituents of true bile are to be met with.

NOTE VIII., CHAPTER II., p. 81.

ANAL RESPIRATION IN CRAYFISH.

Lereboullet ("Note sur une respiration anale observée chez plusieurs Crustacés;" Mémoires de la Société d'Histoire Naturelle de Strasbourg, IV. 1850) has drawn attention to what he terms "anal respiration" in young crayfish, in which he observed water to be alternately taken into and expelled from the rectum fifteen to seventeen times in a minute. I have never been able to observe anything of this kind in the uninjured adult animal, but if the thoracic ganglia are destroyed, a regular rhythmical dilatation and closing of the anal end of the rectum at once sets in, and goes on as long as the hindermost ganglia of the abdomen retain their integrity. I am much disposed to imagine that the rhythmical movement is inhibited, when the uninjured crayfish is held in such a position that the vent can be examined.

NOTE IX. CHAPTER II., p. 82.

THE GREEN GLAND.

The existence of guanin in the green gland rests on the authority of Will and Gorup-Besanez (Gelehrte Anzeigen, d. k. Baienzschen Akademie, No. 233, 1848), who say that in this organ and in the organ of Bojanus of the freshwater mussel, they found "a substance the reactions of which with the greatest probability indicate guanin," but that they had been unable to obtain sufficient material to give decisive results.

Leydig (Lehrbuch der Histologie, p. 467) long ago stated that the green gland consists of a much convoluted tube containing granular cells disposed around a central cavity. Wassiliew ("Ueber die Niere des Flusskrebses:" Zoologischer Anzeiger, I. 1878) supports the same view, giving a full account of the minute structure of the organ, and comparing it with its homologues in the *Copepoda* and *Phyllopoda*.

NOTE X., CHAPTER III., p. 105.

THE ANATOMY OF THE NERVOUS SYSTEM OF THE CRAYFISH.

The details respecting the origin and the distribution of the nerves are intentionally omitted. See the memoir by Lemoine of which the title is given in the "Bibliography."

NOTE XI., CHAPTER III., p. 110.

THE FUNCTIONS OF THE NERVOUS SYSTEM OF THE CRAYFISH.

Mr. J. Ward, in his "Observations on the Physiology of the Nervous System of the Crayfish," (Proceedings of the Royal Society, 1879) has given an account of a number of interesting and important experiments on this subject.

NOTE XII., CHAPTER III. p. 124.

THE THEORY OF MOSAIC VISION.

Oscar Schmidt ("Die Form der Krystalkegel im Arthropoden Auge:" Zeitschrift für Wissenschaftliche Zoologie, XXX. 1878) has pointed out certain difficulties in the way of the universal application of the theory of mosaic vision in its present form, which are well worthy of consideration. I do not think, however, that the substance of the theory is affected by Schmidt's objections.

NOTE XIII., CHAPTER III., p. 135.

THE SPERMATOZOA.

Since the discovery of the spermatozoa of the crayfish in 1835-36 by Henle and von Siebold, the structure and development of these bodies have been repeatedly studied. The latest discussion of the subject is contained in a memoir of Dr. C. Grobben ("Beiträge zur Kenntniss der männlichen Geschlechtsorgane der Dekapoden:" Wien, 1878). There is no doubt that the spermatozoon consists of a flattened or hemispherical body, produced at its circumference into a greater or less number of long tapering curved processes (fig. 34 F). In the interior of this are two structures, one of which occupies the greater part of the body, and, when the latter lies flat, looks like a double ring. This may be called, for distinctness' sake, the *annulate corpuscle.* The other is a much smaller *oval corpuscle,* which lies on one side of the first. The annulate corpuscle is dense, and strongly refracting; the oval corpuscle is soft, and less sharply defined. Dr. Grobben describes the annulate corpuscle as "napfartig," or cup-shaped; closed below, open above, and with the upper edge turned inwards, and applied to the inner side of the wall of the cup. It appeared to me, on the other hand, that the annulate corpuscle is really a hollow ring, somewhat like one of the ring-shaped air-cushions one sees, on a very small scale. Dr. Grobben describes the spermatoblastic cells of the testis and their nuclear spindles; but his account of the development of the spermatozoa does not agree with my own observations, which, so far as they have gone, lead me to infer that the annulate corpuscle of the spermatozoon is the metamorphosed nucleus of the cell from which the spermatozoon is developed. For want of material, however, I was unable to bring my investigations to a satisfactory termination, and I speak with reserve.

Note XIV., Chapter IV., p. 174.

THE MORPHOLOGY OF THE CRAYFISH.

The founder of the morphology of the *Crustacea*, M. Milne Edwards, counts the telson as a somite, and consequently considers that twenty-one somites enter into the composition of the body in the *Podophthalmia*. Moreover, he assigns the anterior seven somites to the head, the middle seven to the thorax, and the hinder seven to the abdomen. There is a tempting aspect of symmetry about this arrangement; but as to the limits of the head, the natural line of demarcation between it and the thorax seems to me to be so clearly indicated between the somite which bears the second maxillæ and that which carries the first maxillipedes in the *Crustacea*, and between the homologous somites in Insects, that I have no hesitation in retaining the grouping which I have for many years adopted. The exact nature of the telson needs to be elucidated, but I can find no ground for regarding it as the homologue of a single somite.

It will be observed that these differences of opinion turn upon questions of grouping and nomenclature. It would make no difference to the general argument if it were admitted that the whole body consists of twenty-one somites and the head of seven.

Note XV., Chapter IV., p. 199.

THE HISTOLOGY OF THE CRAYFISH.

In dealing with the histology of the crayfish I have been obliged to content myself with stating the facts as they appear to me. The discussion of the interpretations put upon these facts by other observers, especially in the case of those tissues, such as muscle, on which there is as yet no complete agreement even as to matters of observation, would require a whole treatise to itself.

Note XVI., Chapter IV., p. 221.

THE DEVELOPMENT OF THE CRAYFISH.

The remark made in the last note applies still more strongly to the history of the development of the crayfish. Notwithstanding the masterly memoir of Rathke, which constitutes the foundation of all our knowledge on this subject; the subsequent investigations of Lereboullet; and the still more recent careful and exhaustive works of Reichenbach and Bobretsky, a great many points require further investigation. In all its most important features I have reason to believe that the account of the process of development given in the text, is correct.

Note XVII., Chapter VI., p. 297.

PARASITES OF CRAYFISHES.

In France and Germany crayfishes (apparently, however, only *A. nobilis*) are infested by parasites, belonging to the genus *Branchiobdella*. These are minute, flattened, vermiform animals, somewhat like small leeches, from one-half to one-third of an inch in length, which attach themselves to the under side of the abdomen (*B. parasitica*), or to the gills (*B. astaci*), and live on the blood and on the eggs of the crayfish. A full account of this parasite, with reference to the literature of the subject, is given by Dormer ("Ueber die Gattung Branchiobdella:" Zeitschrift für Wiss. Zoologie, XV. 1865). According to Gay, a similar parasite is found on the Chilian crayfish. I have never met with it on the English crayfish. The Lobster has a somewhat similar parasite, *Histriobdella*. Girard, in the paper cited in the Bibliography, gives a curious account of the manner in which the little lamellibranchiate mollusk, *Cyclas fontinalis*, shuts the ends of the ambulatory limbs of crayfishes which inhabit the same waters, between its valves, so that the crayfish resembles a cat in walnut shells, and the pinched ends of the limbs become eroded and mutilated.

BIBLIOGRAPHY.

The subjoined list indicates the chief books and memoirs, in addition to those mentioned in the text and in the Appendix, which may be advantageously consulted by any one who wishes to study more fully the biology of the crayfishes.

I.—NATURAL HISTORY.

ROESEL VON ROSENHOF. Der Monatlich-herausgegeben Insekten Belustigung. 1755.

CARBONNIER. L'Écrevisse, Paris, 1869.

BRANDT AND RATZEBURG. Medizinische Zoologie. Bd. II., pp. 58–70.

BELL. British Stalk-eyed Crustacea. 1853.

SOUBEIRAN. Sur l'Histoire naturelle et l'Éducation des Écrevisses. Comptes Rendus, LX., 1865.

CHANTRAN. Observations sur l'Histoire naturelle des Écrevisses. Comptes Rendus, LXXI., 1870.

—— Sur la Fécondation des Écrevisses. Ibid., LXXIV., 1872.

—— Expériences sur la Régénération des Yeux chez les Écrevisses. Ibid., LXXVII., 1873.

—— Observations sur la Formation des Pierres chez les Ecrevisses. Ibid., LXXVIII., 1874.

—— Sur le Mécanisme de la Dissolution intrastomacale des Concrétions gastriques des Écrevisses. Ibid., LXXVIII., 1874.

STEFFENBERG. Bijdrag til kanne domen om flodkraftens natural historia, 1872. Abstract in Zoological Record, IX.

VALLOT. Sur l'Écrevisse fluviatile et sur son parasite l'Astacobdelle branchiale. Comptes Rendus Acad. Sciences, Dijon. Mémoires, 1843–44. Dijon, 1845.

PUTNAM. On some of the Habits of the Blind Crayfish. Proceedings Boston Society of Nat. History, XVIII.

HELLER. Ueber einen Flusskrebs-albino. Verhand d. Z. Bot. Gesellschaft, Wien. Bd. 7, 1857, and Bd. 8, 1858.

LEREBOULLET. Sur les variétés Rouge et Bleue de l'Écrevisse fluviatile. Comptes Rendus, XXXIII., 1857.

GIRARD. Quelques Remarques sur l'Astacus fluviatilis. Ann. Soc. Entom. France, T. VII. 1859.

II.—ANATOMY AND PHYSIOLOGY.

BRANDT AND RATZEBURG. *Op. cit.*

MILNE EDWARDS. Histoire naturelle des Crustacés. 1834.

ROLLESTON. Forms of Animal Life. 1870.

HUXLEY. Manual of the Anatomy of Vertebrated Animals. 1877.

HUXLEY AND MARTIN. Elementary Biology. 1875.

SUCKOW. Anatomisch-Physiologische Untersuchungen. 1818.

KROHN. Verdauungsorgane des Krebses. Gefässsystem des Flusskrebses. Isis, 1834.

VON BAER. Ueber die sogenannte Erneuerung des Magens der Krebse und die Bedeutung der Krebssteine. Müller's Archiv, 1835.

OESTERLEN. Ueber den Magen des Flusskrebses. Müller's Archiv, 1840.

T. J. PARKER. On the Stomach of the Freshwater Crayfish. Journal of Anatomy and Physiology, 1876.

BARTSCH. Die Ernährungs- und Verdauungsorgane des *Astacus leptodactylus*. Budapester Naturhistor. Hefte II. 1878.

DESZÖ. Ueber das Herz des Flusskrebses und des Hummers. Zoologischer Anzeiger, I. 1878.

LEREBOULLET. Note sur une Respiration anale observée chez plusieurs Crustacées. Mém. de la Société d'Histoire Naturelle de Strasbourg, IV., 1850.

WASSILIEW. Ueber die Niere des Flusskrebses. Zoologischer Anzeiger, I. 1878.

LEMOINE. Recherches pour servir à l'histoire des systèmes nerveux, musculaire et glandulaire de l'Écrevisse. Annales des Sciences Naturelles, Sé. IV. T. 15, 1861.

DIETL. Die Organization des Arthropoden Gehirns. Zeitschrift für Wiss. Zoologie, XXVII., 1876.

KRIEGER. Ueber das centrale Nervensystem des Flusskrebses. Zoologischer Anzeiger, I., 1878.

LEYDIG. Das Auge der Gliederthiere. 1864.

MAX SCHULZE. Die Zusammengesetzten Augen der Krebse und Insekten, 1868.

BERGER. Untersuchungen über den Bau des Gehirns und der Retina der Arthropoden. 1878.

GRENACHER. Untersuchungen über das Sehorgan der Arthropoden. 1879.

O. SCHMIDT. Die Form der Krystalkegel im Arthropoden Auge. Zeitschrift für Wiss. Zoologie, XXX., 1878.

FARRE. On the organ of hearing in the Crustacea. Phil. Trans. 1843.

LEYDIG. Ueber Geruchs- und Gehörorgane der Krebse und Insekten. Müller's Archiv, 1860.

HENSEN. Studien über das Gehörorgan der Decapoden. Zeitschrift für Wissenschaftliche Zoologie, XIII. 1863.

GROBBEN. Beiträge zur Kenntniss der männlichen Geschlechtsorgane der Dekapoden. 1878.

BROCCHI. Recherches sur les Organes génitaux mâles des Crustacés décapodes. Annales des Sciences Naturelles, Sé. VI. ii.

LEYDIG. Zur feineren Bau der Arthropoden. Müller's Archiv, 1855.

—— Handbuch der Histologie. 1857.

HAECKEL. Ueber die Gewebe des Flusskrebses. Müller's Archiv, 1857.

BRAUN. Ueber die histologischen Vorgänge bei der Häutung von Astacus fluviatilis. Würzburg Arbeiten, II.

BAUR. Ueber den Bau der Chitinsehne am Kiefer des Flusskrebses und ihr Verhalten beim Schalenwechsel. Reichert u. Du Bois Archiv, 1860.

COSTE. Faits pour servir à l'Histoire de la Fécondation chez les Crustacés. Comptes Rendus, XLVI. 1858.

LEREBOULLET. Recherches sur la mode de Fixation des Œufs aux fausses pattes abdominales dans les Écrevisses. Annales des Sciences Naturelles, Sé. IV. T. 14, 1860.

III.—DEVELOPMENT.

RATHKE. Ueber die Bildung und Entwickelung des Flusskrebses 1829.

LEREBOULLET. Recherches d'Embryologie comparée sur le développement du Brochet, de la Perche et de l'Écrevisse. 1862.

BOBRETSKY. (A Memoir in Russian, of which an abstract is given in Hofmann and Schwalbe, Jahresbericht für 1873 (1875)).

REICHENBACH. Die Embryonanlage und erste Entwickelung des Flusskrebses. Zeitschrift für Wiss. Zoologie. 1877.

IV.—TAXONOMY AND DISTRIBUTION OF CRAYFISHES.

A. *General.*

MILNE EDWARDS. *Op. cit.*

ERICHSON. Uebersicht der Arten der Gattung *Astacus.* Wiegmann's Archiv für Naturgeschichte, XII. 1846.

DANA. Crustacea of the United States Exploring Expedition. 1852.

DE SAUSSURE. Note carcinologique sur la Famille des Thalassinides et sur celle des Astacides. Rev. et Magazin de Zoologie, IX.

HUXLEY. On the Classification and the Distribution of the Crayfishes. Proceedings of the Zoological Society. 1878.

B. *European and Asiatic.*

RATHKE. Zur Fauna der Krym. 1836.

GERSTFELDT and KESSLER. Cited in the text.

DE HAAN. Fauna Japonica. 1850.

LEREBOULLET. Description de deux nouvelles Espèces d'Écrevisses (*A. longicornis, A. pallipes*). Mém. Soc. Science Nat. Strasbourg. V. 1858.

HELLER. Crustaceen des südlichen Europa. 1863.

KESSLER. Ein neuer russischer Flusskrebs, *Astacus colchicus.* Bulletin de la Soc. Imp. des Naturalistes de Moscou, L. 1876.

C. *American.*

STIMPSON. Crustacea and Echinodermata of the Pacific shores of North America. Journal of Boston Society of Natural History, VI.; 1857-8.

DE SAUSSURE. Mémoire sur divers Crustacées nouveaux des Antilles et du Méxique. Mém. de la Société de Physique de Genève T. XIV., 1857.

VON MARTENS. Südbrasilische Süss- und Brackwasser Crustaceen (*A. pilimanus, A. brasiliensis*), Wiegmann's Archiv, XXXV., 1869.

——. Ueber Cubansche Crustaceen. *Ibid.* XXXVIII.

HAGEN. Monograph of the North American *Astacidæ.* 1870.

D. *Madagascar.*

AUDOUIN and MILNE EDWARDS. Sur une Espèce nouvelle du genre Écrevisse (*Astacus*). Écrevisse de Madagascar (*A. Madagascariensis*). Mém. du Muséum d'Hist. naturelle, T. II. 1841.

E. *Australia.*

VON MARTENS. On a new Species of *Astacus*. Annals & Mag. of Natural History, 1866.

HELLER. Reise der "Novara." Zool. Theil. Bd. II. 1865.

F. *New Zealand.*

MIERS. Notes on the Genera *Astacoides* and *Paranephrops*. Transactions of the New Zealand Institute, IX., 1876.

—— *Paranephrops*. Zoology of "Erebus" and "Terror," 1874. Catalogue of New Zealand Crustacea, 1876.

—— Annals of Natural History, 1876.

WOOD-MASON. On the mode in which the Young of the New Zealand *Astacidæ* attach themselves to the Mother. Ann. & Mag. Natural History, 1876.

G. *Fossil Astacomorpha.*

OPPEL. Palæontologische Mittheilungen, 1862.

BELL. British Fossil Crustacea. Palæontographical Society.

P. VAN BENEDEN. Sur la Découverte d'un Homard fossile dans l'Argile de Rupelmonde. Bulletin de l'Acad. Royale de Belgique. XXXIII., 1872.

VON DER MARCK und SCHLÜTER. Neue Fische und Krebse von der Kreide von Westphalen. Palæontologica, XV. 1865.

COPE. On three extinct *Astaci* from the freshwater tertiary of Idaho. Proceedings of the American Philosophical Society, XI, 1869-70.

INDEX.

A.

Abdomen, 19, 141
  development of, 213
Abdominal appendages, 143
  development of, 217
Abdominal somite, characters of, 142
Ætiology, 47
AGASSIZ, 308
Alimentary canal, 51
  development of, 213, 222
Ambulatory legs, 168
American Crayfishes, 243, 247
Amœba, 285
Amurland Crayfishes, 304
Antenna, 23, 172
  development of, 214, 218
Antennule, 23, 173
  development of, 214, 218
Anthrapalæmon, 341
Anus, 29
Apodeme, 99, 158, 175
Appendage, 24, 143, 161, 173
  abdominal, 143
  cephalic, 170
  thoracic, 164
Archenteron, 211
Arctogæal province, 314
Areola, 235
ARISTOTLE, referred to, 4
Arteries, 71
Arteries, development of, 224
Arthrobranchia, 75
Arthrophragm, 158
Arthropoda, 279, 284
Articulations, 95
Asiatic Crayfishes, 304
Astacina, 254
Astacoides, 250, 313
Astacomorpha, 338
Astacopsis, 250, 264
Astacus, division into sub-genera, 290
Astacus angulosus, 302, 310
  colchicus, 302, 310
  dauricus, 304, 310
  fluviatilis,
    anatomy, general account of, 17—31
    attachment of young to mother, 40, 351
    branchial formula, 266
    development, 205–226
    distribution, geographical 44, 288, 298
    distribution, chronological, 44
    ecdysis, 32, 350
    general characters, 6
    growth, 31, 349
    habits, 8

Astacus fluviatilis—continued
histology, 174
mortality, 127
muscular system, 90
myths concerning. 44
name, origin of, 13
nervous system, 101
newly hatched young, characters of, 219
nutrition, 48
occurrence, 5, 8
organs of alimentation, 51
circulation, 68
excretion, 82, 353
hearing, 116
reproduction, 128
respiration, 75, 353
sight, 118
smell, 114
taste, 115
touch, 113
prehension of food, 49
putrid, effect of smell of, 45
reproduction of lost limbs, 38
reproduction, sexual, 39, 128, 135, 350
sexual characters, 7, 20, 32, 145, 241
somites and appendages, 143
systematic description, 230
use as food, 10, 289
varieties, 289
fontinalis, 290
japonicus, 304
klamathensis, 305
leniusculus, 305
Astacus leptodactylus, 299, 302 303, 310, 320
nigrescens, 244
nobilis, 290, 295, 296, 299, 310
oreganus, 305
pachypus, 302, 310
pallipes, 290
politus, 344
saxatilis, 290
Schrenckii, 304, 310
torrentium, 290, 294, 298, 310, 311
tristis, 290
Trowbridgii, 305
Atya, *Atyidæ*, 331, 336
Auditory organ, 116
setæ, 116
Australian Crayfishes, 306
province, 314
Austrocolumbian province, 314
Axius, 271

B.

BALL, R., quoted, 36
Basipodite, 143
BELL, T., quoted, 37, 42
Bile-duct, 61, 66
Biological sciences, scope of, 4
Blastoderm, 207
Blastomere, 205
Blastopore, 209
Blood, 31, 68, 176
corpuscles, 69, 176
development of, 224
sinuses, 50, 69
BOBRETSKY, referred to, 356
BOLIVAR, Dr., 298
Branchiæ,
Astacoides, 266
Astacopsis, 264

Branchiæ—*continued*
Astacus, 25, 75, 265
development of, 224
Cancer, 276
Homarus, 257
Palæmon, 270
Palinurus, 264
Penæus, 267
Branchial chamber, 25
formula,
Astacoides, 266
Astacopsis, 264
Astacus, 266
Cancer, 277
hypothetically complete, 268
Palæmon, 270
Palinurus, 265
Penæus, 267
Branchiobdella, 356
Branchiostegite, 25
development of, 217
BRAUN, quoted, 352
Brazilian Crayfishes, 306

C.

Cæcum, 61
Calcification of exoskeleton, 197
Californian Crayfishes, 243
Cambarus, 44, 247, 310, 312
Cancer, 272, 283
Carapace, 19
development of, 214
CARBONNIER, M., quoted, 297, 349, 350
Cardia, 52
Caridina, 330
Carpopodite, 165
Cell, 66, 199
Cell-aggregate, 190, 199
division, 200
theory, 202, 204
Cephalic appendages, 170
development of, 217
flexure, 163
somites, 154
Cephalon, 19, 141
Cephalothorax, 19
Cervical groove, 19
spines, 234
CHANTRAN, M., quoted, 348, 350, 351
Chelæ, 22
Chilian Crayfishes, 308
Chitin, 50
composition of, 347
Chæraps, 250
Chorology, 46
Circulation, 73
organs of, 68
Common knowledge and science, 3
Connective tissue, 178
development of, 224
COPE, Prof., quoted, 316
Cornea, 118
Coxopodite, 143
Coxopoditic setæ, 78
Crab, see *Cancer*
Crab's-eye, see Gastrolith
Crangon, 272
Crayfish, origin of name, 12
common, see *Astacus fluviatilis*
Crayfishes, Amurland, 304
Asiatic, 304
Australian, 306
Brazilian, 306
Californian, 243
Chilian, 308
definition of, 254
Eastern North American, 247, 305
European, 288, 297
evolution of, 331

Crayfishes, Figian, 306, 313
  Japanese, 304, 313
  Mascarene, 308, 313
  northern and southern, compared, 252
  Novozelanian, 306, 313
  southern, 249
  Tasmanian, 306
  Western North American, 305, 313
Crustacea, 271, 278
Crystalline cones, 121
Cuticle, 33, 50, 175, 192
Cyclas, 356

D.

Dactylopodite, 165
Daphnia, asexual reproduction of 128
DARWIN, C., referred to, 4
DE HAAN, quoted, 313
Development, 205
  abdomen, 213
  abdominal appendages, 217
  alimentary canal, 213, 222
  antennæ, 214, 218
  antennules, 214, 218
  blood and blood vessels, 224
  branchiostegite, 217
  carapace, 214
  cephalic appendages, 217, 219
  connective tissue, 224
  ear, 225
  eye, 225
  eyestalk, 214, 218
  gills, 224
  heart, 224
  kidney, 224
  labrum, 218
  mandibles, 214
Development of muscles, 224
  nervous system, 213, 224
  reproductive organs, 225
  rostrum, 217
  thoracic appendages, 217, 219
Digestion, 63
Distribution, 46
  chronological, of crayfishes, 44, 316, 339
  table of, 345
  geographical, of crayfishes, 44, 288
  causes of, 335
  results of study of, 308, 314
DORMER, quoted, 356
DULK, quoted, 349

E.

Ear, 116
  development of, 225
Ecdysis, 32, 350
Écrevisse à pieds blancs, 289, 297
  à pieds rouges, 289, 297
Ectoderm, 141
Ectostracum, 194
Edelkrebs, 290
Endoderm, 141
Endophragmal system, 157
Endopleurite, 158
Endopodite, 145
Endoskeleton, 17
Endosternite, 158
Endostracum, 194
Engæus, 250, 306
Enoplocytia, 342
Epiblast, 211
Epidermis, 140
Epimeron, 143
Epiostracum, 192
Epipodite, 167

Epistoma, 155
Epithelium, 140, 177
Equus excelsus, occurring with fossil crayfishes, 316
Eryma, 341
Evolution of crayfishes, 331
Excretion, organs of, 82
Exopodite, 145
Exoskeleton, 17
  chemical composition, 347
Eye, 118
  compound, 122
  development of, 225
Eye-stalk, 24, 173
  development of, 214

F.

Family, 252
Fat-cells, 180
Fibre, muscular, 185
Fibril, muscular, 185
Figian Crayfishes, 306
Filament, muscular, 185
Filter of stomach, 58
Flagellum, 167
Food-yelk, 206
Foot-jaws, *see* maxillipedes
Forceps, 22
Foregut, 61
  development of, 213, 222
Fossil crayfishes, 316
FOSTER, Dr. M., referred to, 110
France, consumption of crayfish in, 10
Function, 22

G.

Galaxidæ, 315
Gammarus, 323
Ganglion, 103, 105
Ganglionic corpuscle, 87, 103
Gastric mill, 53
Gastrolith, 29, 347
  chemical composition, 349
Gastrula, 211
GAY, quoted, 356
Genus, 249
Geographical distribution, see Distribution
GERBE, M., quoted, 350
Germinal disc, 209
  layer, 206
  spot, 133
  vesicle, 133
GERSTFELDT, Dr., quoted, 290
Gills, see Branchiæ
GIRARD, quoted, 356
GORUP-BESANEZ, quoted, 353
Green-gland, 83, 353
  development of, 224
GROBBEN, Dr., quoted, 354
Growth of crayfish, 31, 349
Guanin, 82, 353
Gullet, see Œsophagus
GÜNTHER, Dr., quoted, 315

H.

HAGEN, Dr., quoted, 305, 312
Haplochitonidæ, 315
HARVEY, quoted, 5
Head, see Cephalon
Hearing, organ of, 116
Heart, 27, 71
  development of, 224
HELLER, Dr., quoted, 298, 330
Hepatic duct, see Bile duct
Hind gut, 61
  development of, 214, 223
Histology, 175
Histriobdella, 356
Homaridæ, 263

Homarina, 261
Homarus, 13, 42, 257, 332
Homology, homologous, homologue, 148
Hoploparia, 342
Hypoblast, 211

I.

Idothea, 323, 334
Impregnation, 135, 350
Integument, 50
Interseptal zone, 183
Intestine, 29, 61
Ischiopodite, 165

J.

Japanese Crayfishes, 313, 314
Jaws, 23
JOHNSTON, J., quoted, 42

K.

KESSLER, quoted, 298, 304
Kidney, see Green gland
KLUNZINGER, referred to, 330

L.

Labrum, 51
development of, 218
LAMARCK, referred to, 4
LEREBOULLET, quoted, 353
Legs, ambulatory, 168
LEMOINE, referred to, 353
LEYDIG, referred to, 115, 353
Liver, 30, 64
development of, 223
nature of secretion, 352
Lobster, common, see *Homarus*
Norway, see *Nephrops*
Lobster, Rock, see *Palinurus*
LOVÈN, referred to, 327

M.

Machine, living, 128
M'INTOSH, Dr. W. C., quoted, 288
Mandible, 23, 51, 170
development of, 214
MARTENS, VON, 306
Mastodon mirificus, occurring with fossil crayfishes, 316
Maxillæ, 23, 170
Maxillipedes, 23, 164
Medullary groove, 213
Megalopa stage of development, 283
Meropodite, 165
Mesoblast, 212
Mesoderm, 141
Mesophragm, 158
Metamere, 143
Metastoma, 51
Metope, 278
Midgut, 61
development of, 211, 214, 223
MILNE-EDWARDS, quoted, 13, 289
Mollusca, 284
Morphology, 46, 138
comparative, 230
Mortality of crayfishes, 123
Morula, 206
Mosaic vision, 122, 354
Motor plates, 189
Mouth, 51
MÜLLER, JOHANNES, referred to, 122
Muscle, 57, 90, 175, 181
development of, 224
histology of, 90, 181
Muscles of abdomen, 99

Muscles of chela, 93
  of stomach, 57
Myosin, 186
Myotome, 174
Mysis, 281, 323
  relicta, origin of, from *M. oculata*, 327
Mysis stage of development, 280

N.

Natural History, 3
  Philosophy, 3
Nauplius stage of development, 215, 280
Nearctic province, 314
Nephrops, 259, 332
Nerve, 101
  auditory, 117
  optic, 118
Nerve-cells, 103, 187
  fibres, 101, 188
Nervous system, 105
  development of, 213, 224
  functions of, 354
Noble crayfish, see *Astacus nobilis*
Nomenclature, binomial, 13, 15
Norway lobster, see *Nephrops*
Novozelanian province, 314
Nucleated cell, 199
Nucleolus, 187
Nucleus, 177, 200
  changes of, in cell-division, 200

O.

Œsophagus, 51
Olfactory organ, 114
Organ, 22
Origin of crayfish, evidence as to, 320, 331
Ovary, 31, 129
  structure of, 131
Oviduct, 129
Oviposition, 351
Ovisac, 132
Ovum, 129
  structure of, 133

P.

Palæarctic province, 314
Palæmon, 268, 328
Palinuridæ 263
Palinurus, 261, 264
Palp, 171
Paranephrops, 250, 306, 313
Paraphragm, 158
Parasites of crayfish, 356
Parastacidæ, 252, 256, 306, 313
Parastacus, 250, 306
Pemphix, 341
Penæus, 267, 280
Pericardium, 69
Perivisceral cavity, 50
Phyllobranchia, 271
Physiology, 46
Pleurobranchia, 79
Pleuron, 96, 143
Podobranchia, 75, 165
Podophthalmia, 279
Pore-canals, 195
Post-orbital ridge, 233
  spine, 232
Potamobiidæ, 252, 256
Prawn, see *Palæmon*
Prehension of food, 49
Procephalic lobes, 160
  development of, 213
Propodite, 165
Protopodite, 143
Prototroctes, 315

Protozoa, 285
Pseudastacus, 343
Pylorus, 52

R.

Race, 292
RATHKE, quoted, 356
RÉAUMUR, quoted, 33
Reflex action, 108
REICHENBACH, quoted, 356
Renal organ, see Green-gland
Reproduction of lost limbs, 38
sexual, 39, 128, 135, 350
Reproductive organs, 128
development of, 225
Respiration, anal, 353
Respiratory organs, see Branchiæ
Retropinna, 315
ROBIN, quoted, 352
Rock lobster, see *Palinurus*
ROESEL VON ROSENHOF, quoted, 41, 43
RONDOLETIUS, referred to, 4
Rostrum, 157
development of, 217

S.

Salivary glands, 352
Salmonidæ, parallel between their distribution, and that of *Astacidæ*, 315
Sarcolemma, 90, 182
SARS, G. O., referred to, 327
SARTORIUS VON WALTERHAUSEN, quoted, 322
Scaphognathite, 80, 170
Schizopod stage of development, 280
SCHLÜTER, 317
SCHMIDT, O., quoted, 354
SCHRANK, 290
Science, physical, 3
Science and common sense, 1
Segmentation, 174
Self-causation, 112
Sensory organs, 113
Septal line, 183
zone, 183
Setæ, 197
Shrimp, see *Crangon*
SIEBOLD, VON, referred to, 331
Sight, organ of, 118
Sinus, sternal, 69
Smell, organ of, 114
Somite, 143, 161, 355
abdominal, 142
cephalic, 154
thoracic, 150
SOUBEIRAN, M., quoted, 349
Southern Crayfishes, 249
Species, 243, 290
morphological, 291
physiological, 296
Spermatozoa, 129, 135, 354
Spontaneous action, 112
Squame of antenna, 172
Steinkrebs, see *Astacus torrentium*
Sternum, 96, 143
Stomach, 29, 51
Stone-crayfish, see *Astacus torrentium*
Striated spindle, 121
Swimmeret, 20

T.

Taste, organ of, 115
Teleology, 47, 137
Tendon, 92, 175

Tergum, 96, 143
Terminal plates, 189
Terminology, scientific, 14
Testis, 129
 structure of, 133
Thoracic appendages, 164
 development of, 217
 somites, 150
Thorax, 19, 141
Tissue, 175
Touch, organ of, 113
Transformism, 318
TREVIRANUS, referred to, 4
Tribe, 252
Trichobranchiæ, 263
Troglocaris, 337

V.

Valves of heart, 73
 of stomach, 59
VAN HELMONT, quoted, 45
Variety, 290, 292
Vas deferens, 130
Vent, see Anus
Vertebrata, 284
Vertebrata, eye of, 122, 125
Visual pyramid, 121
 rod, 121
Vitelline membrane, 133
Vitellus, 133
Voluntary action, 112
VON DER MARCK, 317

W.

WARD, J., referred to, 354
WASSILIEW, quoted, 353
Whirlpool of life, 84
WILL, quoted, 353
WOOD-MASON, quoted, 44

Y.

Yelk, 133
Yelk-division, 205
Young of *Astacus*, newly hatched, characters of, 219

Z.

Zoæa stage of development, 280

THE END.

V.

Solar Light and Heat:

THE SOURCE AND THE SUPPLY. Gravitation: with Explanations of Planetary and Molecular Forces. By ZACHARIAH ALLEN, LL. D. With Illustrations. 1 vol., 8vo. Cloth. Price, $1.50.

VI.

Macaulay's Essays.

ESSAYS CRITICAL AND MISCELLANEOUS. By Lord MACAULAY. In two vols., 8vo. Cloth. Price, $2.50.

This is a remarkably cheap edition of Macaulay's Essays. It is printed in good style, and handsomely bound.

VII.

A Class-Book History of England.

Illustrated with numerous Woodcuts and Historical Maps. Compiled for Pupils preparing for the Oxford and Cambridge Local Examinations, and for the Higher Classes of Elementary Schools. By the Rev. DAVID MORRIS, Classical Master in Liverpool College. First American from fifteenth English edition. 1 vol., 12mo. Cloth. Price, $1.25.

VIII.

The English Language,

AND ITS EARLY LITERATURE. By J. H. GILMORE, A. M., Professor of Logic, Rhetoric, and English, in the University of Rochester. 1 vol., 12mo. Cloth. Price, 60 cents.

IX.

DR. RICHARDSON'S

"Ministry of Health."

A Ministry of Health, and other Addresses. By BENJAMIN WARD RICHARDSON, author of "Diseases of Modern Life." 1 vol., 12mo. Cloth. Price, $1.50.

X.

Harvey's Therapeutics.

First Lines of Therapeutics, as based on the Modes and the Processes of Healing, as occurring spontaneously in Disease, etc., etc. By ALEXANDER HARVEY, M. D., Professor of Materia Medica in the University of Aberdeen. 1 vol., 12mo. Cloth. Price, $1.50.

For sale by all booksellers; or any work sent, prepaid, to any address in the United States, on receipt of the price.

D. APPLETON & CO., Publishers, New York.

Published by

Facsimile Publisher
Ashok Vihar, Delhi-110052 (India)
Website: www.facsimilepublisher.com

Distributed by

Gyan Books Pvt. Ltd.
5, Ansari Road, Daryaganj, New Delhi-110002 (India)
E-mail: books@gyanbooks.com, Website: www.gyanbooks.com

2 222000 008208